Analytical Modeling in Applied Electromagnetics

Sergei Tretyakov

Artech House
Boston • London
www.artechhouse.com

Library of Congress Cataloging-in-Publication Data
Tretyakov, Sergei.
  Analytical modeling in applied electromagnetics / Sergei Tretyakov.
    p. cm. — (Artech House Artech House electromagentic analysis series)
  Includes bibliographical references and index.
    ISBN 1-58053-367-1 (alk. paper)
      1. Electromagnetism—Mathematics. 2. Electromagnetic fields—Mathematical models.
  3. Magnetic materials—Mathematical models. I. Title. II. Series.

QC760.T74 2000
537'.01'51—dc21

                                                                        2003048094

British Library Cataloguing in Publication Data
Tretyakov, Sergei
  Analytical modeling in applied electromagnetics. — (Artech
  House electromagentic analysis series)
  1. Materials—Electric properties 2. Materials—Analysis
  3. Electromagnetism—Mathematical models
  I. Title
  620.1'1297

  ISBN 1-58053-367-1

Cover design by Igor Valdman

© 2003 ARTECH HOUSE, INC.
685 Canton Street
Norwood, MA 02062

International Standard Book Number: 1-58053-367-1
Library of Congress Catalog Card Number: 2003048094

10 9 8 7 6 5 4 3 2 1

Contents

Preface

This book is about understanding and analytical modeling of material layers, composites, artificial materials, and metamaterials. It explains electromagnetic properties of new artificial electromagnetic surfaces and metamaterials, especially size-dependent phenomena in materials; materials with predefined, engineered, and electrically controllable properties; backward-wave, double negative, or "left-handed" materials; and novel prospective devices such as the perfect lens.

Strictly speaking, no realistic electromagnetic problem related to radio engineering can be solved by brute force, just based on the microscopic Maxwell equations: there are simply too many molecules in any macroscopic device of practical interest. This forces us to introduce *models*, defining and estimating permittivity, conductivity, and other effective parameters of materials. The various models that we use in applied electromagnetics form a kind of hierarchical structure. Starting from the microscopic Maxwell equations we introduce macroscopic models of materials. Next, we need to solve problems where we have, for example, thin layers of various materials. At this stage, assuming the medium parameters are known, we develop models for thin layers. At a still higher level, we study, for example, periodical structures formed from layers of certain materials or metamaterials with complex and engineered inclusions. Here we need models describing the electromagnetic responses of these systems assuming the properties of individual layers or inclusions to be known.

This book is designed to provide working examples of approximate analytical models used in electromagnetics, especially for new applications. First, it describes basic elements that are building blocks of every electromagnetic structure (small particles, composite materials, layers of various materials, artificial materials, and so on), continuing with models of complex and exotic metamaterials. The analysis includes classical and well-studied topics such as highly conducting surfaces and thin layers of various materials. On the other hand, novel structures are explained, including artificial impedance surfaces, electromagnetic (photonic) bandgap structures, metamaterials, and artificial materials with negative parameters (backward-wave materials or "left-handed" media). Explaining classical topics, we build models that describe various structures in a unified way and present novel or alternative derivations. Considering new materials, the emphasis is on analytical models of complicated structures, which are usually analyzed by pure numerical means.

The book can serve both as a tutorial presenting new developments in the field of periodical structures, electromagnetic composites, and metamaterials, and as a textbook for postgraduate students providing simple and useful models for materials and components needed in many current applications. Examples show the use of models in analytical and also numerical modeling of practical devices, especially antennas, microwave components, and radar absorbers with the use of advanced artificial materials.

Most of the book's content is based on my own work and the work of colleagues from my research group. Our research interests naturally affected the choice of topics. Compiling the reference lists, I tried to include references to the latest results of other authors, as well as to review papers and monographs, to help the reader in his or her work. Naturally, such reference lists cannot be complete, especially in a book on a broad topic. Research work reflected in this book has been done in cooperation with many colleagues from Helsinki University of Technology (Finland), St. Petersburg State Technical University (Russia), and from other research centers. I feel grateful for their contributions and help. My students helped a lot, especially P.A. Belov, M.K. Kärkkäinen, and S.I. Maslovski. Special thanks go to I.T. Rekanos, C.R. Simovski, and A.P. Vinogradov, who read parts of the manuscript and gave valuable recommendations.

Chapter 1

Introduction

The role of analytical modeling in radio engineering is twofold. To develop more and more advanced devices for more and more versatile applications, we need to study more and more complex electromagnetic systems. Models simplify (or even make possible) solutions needed in the practical design tasks. Most often this means that some unimportant features of the system are discarded so that the description is simplified. On the other hand, we need deep physical understanding of electromagnetic behavior of rather complicated structures utilized in the design of new devices. One can even say that a good researcher should "feel" how a certain system will behave before making any numerical simulations or analytical estimations. Here, analytical models are of fundamental and principal importance. Deep understanding of physical phenomena in materials and structures is a necessary prerequisite for inventions of new designs and new devices.

One might think that approximate models in science play a kind of secondary and auxiliary role. However, it is not so (an interesting discussion on this subject can be found in [1]). Actually, all equations and solutions in physics are approximate: Even the most general equations of theoretical physics are models that do not account for still unknown effects. One of the physical models, the very simple approximate model of the classical Maxwell equations, does not include, for instance, quantum effects. The fundamental role of the approximate nature of electromagnetic equations can be understood from the fact that only approximations in our modeling allow us to introduce meaningful physical values and parameters, even such fundamental notions as the electric charge and electromagnetic field. Taking as a more practical example the analytical modeling of materials, we observe that only after making approximations can we introduce such physical parameters as the permittivity and permeability.

In the title of the book the emphasis is on *modeling*. Here, we do not present known analytical solutions of canonical problems of electromagnetics. Rather, we build approximate *models* using analytical means that allow analytical solutions of quite involved problems. This book is mainly oriented towards applied electromagnetics, so our models and parameters resulting from approximations will be directly applicable to practical tasks (effective permittivity, grid impedance, surface impedance, transfer matrix, and so forth).

1.1 BACKGROUND AND APPLICATIONS

We will start with modeling of thin layers. From the point of view of applications, this is related to the calculation of the scattering from arbitrary bodies, especially bodies that are large comparable to the wavelength. For example, for radar applications we need to estimate radar signatures of aircrafts, ships, masts, and so forth [2]. Another application area is in radar absorbers and the evaluation of their performance for complex-shaped bodies covered by absorbing layers. Usually, to minimize the radar cross section for a specific direction or directions, the shape of the target is optimized, and then the most critical parts of the body are covered by a thin absorbing coating. The calculation of fields reflected by complex-shape metal bodies with complex coatings is a very demanding task and even with the use of supercomputers cannot be solved without the use of approximate models. A material body with a coating and its model are illustrated in Figure 1.1. The electromagnetic properties of the body and the coating are modeled by an impedance relation between the tangential electric and magnetic fields at the coating surface. If the approximate surface impedance is a simple operator (or even a scalar), the calculation is greatly simplified.

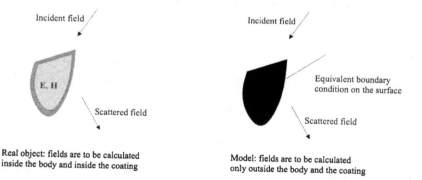

Figure 1.1 A metal body with a coating of another material and its model. In the model formulation, inside the body (represented by a "black box") there are no fields.

Models of thin layers of conductors and dielectrics are also needed in the analysis of microwave circuits. For example, accurate calculation of the loss factor due to imperfect conductivity in microstrip components by a direct numerical solution of the appropriate electromagnetic problem is usually not realistic due to very quickly varying fields inside metal. Approximate analytical models that replace a metal layer by a surface impedance boundary condition (possibly, by a higher-order condition) dramatically simplify computations, and in some cases even allow analytical solutions of fairly complicated problems.

If a layer is thin in wavelengths, the field components tangential to the interfaces cannot change much across the layer. This suggests that we can average these fields inside the layer. A lot of unimportant information is dropped out: Instead of considering the field distribution at every point inside the slab, one only retains the averaged values, dramatically reducing the number of unknowns in the problem. In this sense this method is a good introduction into effective media models of materials (Chapter 5), since in the modeling of materials the goal is also to reduce the complexity introducing averaged macroscopic fields and describing the material with just a few effective parameters, usually the permittivity and permeability. For a thin material layer on a metal surface, for example, a similar role is played by the equivalent surface impedance. For a layer sandwiched between two other material regions, components of the transmission matrix connecting the fields at the layer interfaces are sufficient for approximate description of the layer properties.

It appears obvious that approximate models are very much needed for calculation of reflection from bodies with coverings or surface treatments, but in fact even if the body is a good conductor with no covering at all, we actually use approximate models replacing the body by an approximate surface impedance. The simplest approximation is the ideal electric wall boundary condition. To account for losses, the simple impedance boundary condition is usually applied. Often it is assumed that this is a very good approximation for radar cross section calculations, but if the surface is very curved or the external field varies very fast along the surface, this simple method fails. More accurate higher-order boundary conditions should be used in this case.

The simplest impedance boundary condition has been known since the 1940s [3, 4] (a historical survey can be found in [5]). This model allowed analytical solutions of many practical problems of radio wave propagation (in many situations the Earth's surface can be modeled by a simple impedance boundary condition), scattering by large targets, microwave techniques (e.g., estimation of waveguide losses due to finite conductivity of walls). The success of the model motivated many extensions and generalizations. Higher-order impedance boundary conditions, also called generalized impedance boundary conditions, have been developed by many authors (see monographs [6, 7]). In this book, models are also developed for multilayered and inhomogeneous coatings, and applications in numerical techniques are also shown. An imaginary interface between two regions of free space can be replaced by an operational impedance boundary condition, thus eliminating the need to compute fields in one of the regions. An appropriate simplification of the impedance operator leads, after a discretization, to an absorbing boundary condition.

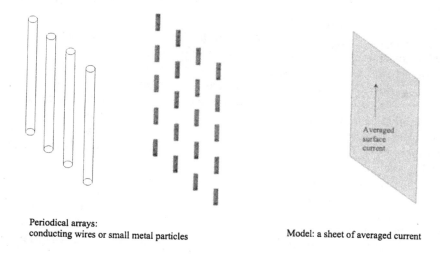

Periodical arrays:
conducting wires or small metal particles

Model: a sheet of averaged current

Figure 1.2 If the period is small compared to the wavelength, an array with a complicated distribution of current on conductors and a complex field structure near the grid can be replaced by an averaged current sheet.

Next we consider modeling of various arrays and meshes. There are two main application areas for periodical structures in radio engineering. First, they serve as slow-wave structures (e.g., in microwave tubes [8]). Second, stopbands are exploited in filters. In the optical region, periodical structures with stopbands for electromagnetic waves are called *photonic bandgap materials*, and stopbands are called *bandgaps*. These novel materials find important applications in laser technology and, most recently, telecommunication applications have been suggested. In addition, metal grids are used as screens and antenna reflectors. First, we study dense arrays (Chapter 4): The period is supposed to be small with respect to the wavelength. A typical application is a metal grid antenna reflector. In the design of grid reflectors we need, first of all, to estimate the reflection coefficient, which should be close to −1. To predict side and back-lobe levels, we have to solve the diffraction problem for the whole grid, taking into account inhomogeneities of the current distribution near the reflector edge. In practical antennas, the number of individual wires forming the grid reflector is usually very large, so the calculation of currents induced on individual wires is not realistic and not reasonable. However, at distances larger than the grid period, the reflected field is already rather uniform, and can be considered as created by the average current on the reflector (Figure 1.2). This is the main idea behind the averaged boundary conditions for periodical arrays of wires [9] or arrays of small scatterers. Approximate conditions are established for the averaged fields in the array plane. Using these conditions, complicated diffraction problems can be tackled even by analytical means.

When the period of the grid or (and) the inclusion size is comparable to the wavelength, resonances occur, and such structures are widely used as frequency selective surfaces. They are usually analyzed numerically, mainly with the periodical method of moments, and we do not cover this regime in this book about analytical methods, referring instead to existing monographs [10, 11].

Considering models of materials, our main interest here will be in new, artificial, complex, or metamaterials [12, 13]. Most often this means that the particles forming a composite material are designed for a specific application. They can have complicated forms or even electronic circuits inside. This way it becomes possible to realize "materials" with properties not found in any natural material or in any mixture of different natural materials. If the inclusions are active (meaning that they contain power sources or some power is supplied to the inclusions), most exotic desires of an engineer regarding electromagnetic properties of materials can be satisfied in principle. In this book, we build models of metamaterials step by step. To mimic natural materials, artificial particles should react to external fields like natural molecules, that is, the fields should be able to induce electric and magnetic dipole moments in the inclusions. We can choose simple small electric and magnetic dipole antennas as artificial inclusions, because there is a possibility to modify the response of these particles loading the antennas by bulk loads (possibly even by active loads or electronic circuits). To understand complex metamaterials, the first step is to study single individual inclusions and determine polarizabilities of small loaded dipole scatterers.

The next step is modeling of the effective medium performance. Two cases must be distinguished. At low frequencies, the whole three-dimensional array or mixture can be modeled as a homogeneous effective medium (Figure 1.3) with the help of various mixture rules [14]. As the frequency increases, the wavelength becomes comparable with the inhomogeneity scale, and the description in terms of effective parameters of an equivalent *homogeneous* medium loses its accuracy and physical meaning. We say that the medium is spatially dispersive. If the spatial dispersion effects are weak (the wavelength is still considerably longer than the array period), the effective medium description is still possible, but the constitutive relations of the effective-medium model are more complicated, and more parameters are needed for an adequate description of the medium properties [15, 16]. The first-order effects of spatial dispersion can be modeled by the chirality parameter [17].

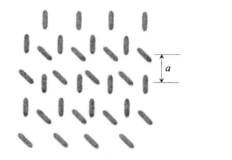

Real system: 3D array of particles

Model for $a \ll \lambda$: effective medium

Figure 1.3 If the distance between particles is small compared to the wavelength, a three-dimensional array or a random mixture of particles can be modeled by an effective medium. If this distance or (and) the particle size is comparable or larger than the wavelength, we model the system by its dispersion equation.

Periodical materials with strong spatial dispersion are called *photonic crystals* or *electromagnetic crystals* or *photonic (electromagnetic) bandgap structures* [18–21]. Researchers and engineers working with microwave applications are familiar with electromagnetic wave phenomena in one- and two-dimensional periodical structures. Periodical structures in waveguides are used in the design of filters and traveling-wave amplifiers and generators (e.g., [8]). Two-dimensional, most commonly planar, periodical structures we find in frequency selective surfaces and phased antenna arrays [10, 11, 22]. Three-dimensional periodical structures have been actively studied since the late 1980s. The main phenomena are similar to those known at lower dimensions: Frequency stopbands (bandgaps) exist, such that no propagating modes exist within these frequency bands. Naturally, the spectrum is different for waves traveling along different directions and having different polarizations. The existence of a *complete* stopband in three-dimensional periodical structures, such that *all* electromagnetic modes are evanescent within this frequency band, was theoretically predicted [23] in 1990 and confirmed experimentally [24] in 1991. In optics, this kind of material is perhaps the ideal candidate for the laser medium, because the spontaneous emission is suppressed within the stopband. In the microwave region, there are many various applications, and potentials are very promising. These new materials extend our possibilities in the design of antennas and high-speed integrated circuits. For example, it is possible to suppress surface waves along substrates of microstrip antennas and in phased arrays. In mobile communication technologies, these materials allow better control of the near-field distribution of mobile terminal antennas, thus helping to reduce unwanted electromagnetic interactions with the user. New waveguiding structures with extremely low reflections at bends have been proposed.

Only some simple three-dimensional electromagnetic crystals allow analytical solutions (Chapter 5), but these results are very important for understanding of the physical phenomena in more complex structures. In this book, we explain the main concepts needed for research work in this new area and provide working models for various practical crystals.

Still more general and versatile materials can be built if the constituents are especially designed with the use of bulk loads and electronic circuits. It appears to be possible to construct a very general analytical model for these metamaterials that gives the design criteria for specific applications (Chapter 6). In order to design a material with the required properties, we can use, for example, arrays of small loaded antennas or arrays of loaded wires. The analytical methods developed here allow us to predict the metamaterial properties as dependent on the inclusion geometry and load properties.

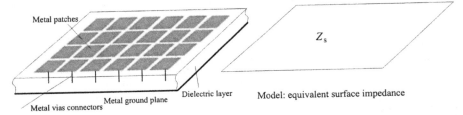

Metal patches

Z_s

Metal vias connectors   Metal ground plane   Dielectric layer   Model: equivalent surface impedance

Figure 1.4 If the array period is small compared to the wavelength, the whole complicated system can be modeled by an equivalent impedance surface.

Also two-dimensional "metasurfaces" can be applied in antenna and absorber technologies. Dense arrays of metal patches parallel to the ground plane and connected to the ground by vias wires are modeled by a resonant reactive surface impedance. These *mushroom structures* [25] find applications in antennas and microwave devices mainly because they do not support surface waves within a certain frequency range and allow us to realize coverings with the desired surface impedance. Layers that are thin compared to the wavelength inside the layer but still reacting very strongly to incident electromagnetic waves are called *artificial high-impedance surfaces*. The mushroom layer is one example of such surfaces. A simple model of an artificial impedance surface is illustrated in Figure 1.4.

One specific example of exotic artificial materials is the medium with negative real parts of the effective permittivity and permeability [26]. The fact that this medium has many names (*Veselago medium, backward-wave medium, left-handed medium, double negative material*) shows the degree of interest and potential of new materials with these properties. One of the suggested applications is the perfect lens [27]: It appears that a slab of Veselago medium can *amplify* evanescent modes of external sources. Explanations and models of these phenomena are given in Chapter 6.

1.2   ORGANIZATION OF THE BOOK

It was not easy to decide what would be the most logical structure of this book. One possibility would be to treat models of materials first, because this is the first step in developing complicated models from the microscopic Maxwell equations. However, this step is perhaps one of the most difficult in electromagnetics, and much knowledge of other models is needed to understand materials. That is why the book starts with building models of thin material layers, assuming at this stage that the properties of materials forming the layers are known. Models based on averaging of the field inside the layer are introduced for thin layers of various materials (e.g., dielectrics and ferrites). Many of the general concepts introduced in Chapter 2 will be used in the following chapters.

Naturally, the question arises: How accurate is the model? Can we somehow estimate the error and learn when the model becomes useless? Obviously, no well-founded answer can be given to such questions unless a more accurate model or an exact solution is available. Fortunately, in this particular case of a planar isotropic layer the exact solution for the field distribution exists. In Chapter 3 we derive exact counterparts of the approximate representations for material layers, and clarify the physical meaning of the approximations. Consideration of this particular example also gives a good opportunity to discuss approximate models in a more general case. The exact solution for a slab is next used to introduce more accurate higher-order equivalent boundary and sheet conditions for layers and media interfaces (Chapter 3). These *analytical* models can be used also in *numerical* techniques. Here, the analytical models of an interface with an isotropic half space are used to introduce a class of absorbing boundary conditions for termination on finite-difference time-domain computational lattices. At the end of Chapter 3, material layers covering ideal and real metal bodies are discussed.

The next chapter is devoted to periodical structures and dense meshes, and it starts with an introduction of the main concepts, such as Floquet modes. In this chapter the emphasis is on arrays and meshes with small periods, although the general case is also considered in detail. However, the main feature that allows simplifications of models is the smallness of the array period or of the particle size. For example, wire grids can be replaced by an equivalent current sheet, and modeled by transition conditions for the tangential fields. A similar simplification is possible for periodical arrays of small particles, and the introduced models are applicable even for the array periods comparable with the wavelength.

To become equipped for the design of artificial composite materials, we first need to understand electromagnetic properties of artificial molecules, which can be small metal particles of various shapes. In particular, in Chapter 5 we introduce analytical formulas for the polarizabilities of short dipole and wire antennas loaded by arbitrary loads. This general approach opens a way to design rather general metamaterials with required properties and even electrically controllable and adaptive materials. This model is the basis of the general theory of metamaterials introduced in Chapter 6. As soon as the properties of small inclusions are understood, we continue with the modeling of materials as such. Both random and periodical structures are studied. Chapter 5 explains artificial dielectrics and electromagnetic (photonic) crystals using as simple as possible analytical approaches.

Chapter 6 is devoted to new applications, especially in so-called *metamaterials*, that is, artificially created materials and surfaces with the use of active or controllable inclusions. First, we consider the problem of material synthesis in a rather general formulation. Suppose you want to design a medium with a certain dispersion law of the effective permittivity. Using small loaded wire inclusions, what should be the loads? What is possible and what is not? Artificial impedance surfaces (sometimes also called *high-impedance surfaces*) are considered next. What are the factors that determine the surface properties? How do we design the surface with desired properties? How does the surface react at obliquely incident waves? These questions are addressed here. The last topic of the book is the modeling of new artificial materials with negative real parts of the effective permittivity and permeability. We start from a general discussion of the wave phenomena in materials with negative parameters. These media exhibit unusual properties like support of backward waves (when the energy flow of a time-harmonic plane wave is directed opposite with respect to the wave vector) and anomalous negative refraction of plane monochromatic electromagnetic waves. Perhaps the most exciting application, the perfect lens which is in principle capable of focusing not only propagating but also evanescent waves, is analyzed and the performance limitations are discussed. Various possible realizations using the concept of metamaterials are discussed, using the models introduced in the book.

References

[1] Fok, V.A., "On the Principal Importance of Approximate Methods in Theoretical Physics," *Uspekhi Fizicheskikh Nauk,* Vol. 16, No. 8, 1936, pp. 1071-1083 (in Russian).

[2] Knott, E.F., J.F. Shaeffer, and M.T. Tuley, *Radar Cross Section,* Norwood, MA: Artech House, 1993.

[3] Shchukin, A.N., *Propagation of Radio Waves, Part I: Basic Theory of Wave Propagation over a Plane and a Sphere*, Moscow-Leningrad: Navy Publishers, USSR, 1940 (in Russian).

[4] Leontovich, M.A., *Investigations on Radio Wave Propagation, Part II*, Moscow: Academy of Sciences, 1948 (in Russian).

[5] Pelosi, G., and P.Ya. Ufimtsev, "The Impedance-Boundary Condition," *IEEE Antennas and Propagation Magazine*, Vol. 38, No. 1, 1996, pp. 31-35.

[6] Senior, T.B.A., and J.L. Volakis, *Approximate Boundary Conditions in Electromagnetics*, London: The Institute of Electrical Engineers, 1995.

[7] Hoppe, D.J., and Y. Rahmat-Samii, *Impedance Boundary Conditions in Electromagnetics*, Washington, D.C.: Taylor & Francis, 1995.

[8] Hutter, R.G.E., *Beam and Wave Electronics in Microwave Tubes*, Princeton, NJ: Van Nostrand, 1960.

[9] Kontorovich, M.I., et al., *Electrodynamics of Grid Structures*, Moscow: Radio i Svyaz, 1987 (in Russian).

[10] Munk, B.A., *Frequency Selective Surfaces: Theory and Design*, New York: John Wiley & Sons, 2000.

[11] Wu, T.K., (ed.), *Frequency Selective Surface and Grid Array*, New York: John Wiley & Sons, 1995.

[12] Zouhdi, S., A. Sihvola, and M. Arsalane, (eds.), *Advances in Electromagnetics of Complex Media and Metamaterials*, NATO Science Series: II: Mathematics, Physics, and Chemistry, Vol. 89, Dordrecht: Kluwer Academic Publishers, 2003.

[13] Ziolkowski, R.W., and N. Engheta, (eds.), Special Issue on Metamaterials, *IEEE Trans. Antennas and Propagation*, Vol. 51, 2003.

[14] Sihvola, A., *Electromagnetic Mixing Formulas and Applications*, London: The Institute of Electrical Engineers, 1999.

[15] Serdyukov, A.N., et al., *Electromagnetics of Bi-Anisotropic Materials: Theory and Applications*, Amsterdam: Gordon and Breach Science Publishers, 2001.

[16] Vinogradov, A.P., *Electromagnetics of Composite Materials*, Moscow: URSS, 2001 (in Russian).

[17] Lindell, I.V., et al., *Electromagnetic Waves in Chiral and Bi-Isotropic Media*, Norwood, MA: Artech House, 1994.

[18] Joannopoulos, J.D., et al., *Photonic Crystals,* Princeton, NJ: Princeton University Press, 1995.

[19] Sakoda, K., *Optical Properties of Photonic Crystals,* Springer Series in Optical Sciences, Vol. 80, New York: Springer-Verlag, 2001.

[20] Mini-special Issue on Electromagnetic Crystal Structures, Design, Synthesis, and Applications, *IEEE Trans. Microwave Theory and Techniques,* Vol. 47, No. 11, 1999.

[21] Feature Section on Photonic Crystal Structures and Applications, *IEEE J. Quantum Electronics,* Vol. 38, No. 7, 2002.

[22] Hansen, R.C., *Phased Antenna Arrays,* New York: John Wiley & Sons, 1998.

[23] Ho, M., C.T. Chan, and C.M. Soukoulis, "Existence of a Photonic Gap in Periodic Dielectric Structures," *Physical Review Lett.,* Vol. 65, 1990, pp. 3152-3155.

[24] Yablonovich, E., T.J. Gmitter, and K.M. Leung, "Photonic Band Structure: The Face-Centered-Cubic Case Employing Nonspherical Atoms," *Physical Review Lett.,* Vol. 67, 1991, pp. 2295-2298.

[25] Sievenpiper, D., et al., "High-Impedance Electromagnetic Surfaces with a Forbidden Frequency Band," *IEEE Trans. Microwave Theory Techniques,* Vol. 47, No. 11, 1999, pp. 2059-2074.

[26] Veselago, V.G. "The Electrodynamics of Substances with Simultaneously Negative Values of ϵ and μ," *Soviet Physics Uspekhi,* Vol. 10, 1968, pp. 509-514 (originally pubished in Russian in *Uspekhi Fizicheskikh Nauk,* Vol. 92, 1967, pp. 517-526).

[27] Pendry, J.B., "Negative Refraction Makes a Perfect Lens," *Physical Review Lett.,* Vol. 85, 2000, pp. 3966-3969.

Chapter 2

Thin Layers and Sheets

In this chapter, we will deal with planar or slightly curved material layers. Solutions of many practical problems can be dramatically simplified, if we can design such a model that takes into account the properties of the layer by some effective boundary conditions. Such conditions connect the fields only on the boundaries of the slab or even on some effective surfaces, so that the field distribution inside the layer is no longer necessary to be considered. The effective boundary conditions approach is most effective for thin layers: The field distribution is difficult to evaluate numerically (very small cells are needed), but in this particular case the boundary condition approximation is simple and very accurate. The situation is similar to the case of a simple interface between two media. We can explicitly take into account the existence of a thin transition region and solve the Maxwell equations in this inhomogeneous structure without using any boundary conditions, or we can make use of the model of an abrupt change of the medium properties and use the Maxwellian boundary conditions of continuity of the tangential field components. If the transition region is extremely thin, the accurate calculation of the field distribution becomes very difficult, but the simple boundary condition becomes very accurate precisely in the same situation. The use of effective boundary conditions for thin layers is an old method, and references to classical papers can be found in monographs [1, 2]. Our approach to this problem is based on the field averaging and on the use of the exact solution for planar layers.

2.1 INTRODUCTORY NOTES

One can say that the effective boundary conditions approach means splitting the problem into two steps: First, we analyze the fields only *inside* the slab and find appropriate boundary conditions; next, we study the fields only *outside* the layer. Of course, there is no need to repeat the first step for every

13

particular problem since the boundary conditions have been determined.

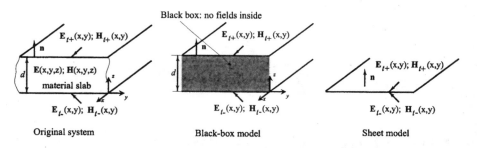

Figure 2.1 Two possible models of a thin layer: a black box of a finite thickness and a sheet of zero thickness.

It is important to note that since the fields inside the material slab have been excluded from the analysis, what is left can be either represented as a "black box" of a finite thickness (naturally equal to the physical layer thickness) or by an effective *sheet*; that is, a surface of zero thickness. This is illustrated in Figure 2.1 for a material slab. The choice of the preferable model depends on the particular application.

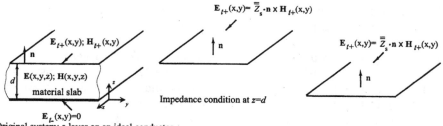

Figure 2.2 Two possible models of a thin layer on a metal surface: an impedance boundary condition at $z = d$ (there are no fields at $z < d$) and an impedance boundary condition at $z = 0$ (there are no fields at $z < 0$). Impedance operators $\overline{\overline{Z}}_s$ are different for the two models.

Similarly, for a thin layer covering an ideally conducting surface or any boundary impenetrable to electromagnetic fields, it is possible to formulate equivalent approximate boundary conditions either at the free surface of the layer or just on the boundary surface (see Figure 2.2). The difference between the two models is evident from the analysis of the limiting case when the properties of the layer become the same as that of the space above the structure. The boundary condition formulated on the conductor surface reduces to the trivial condition $\mathbf{E}_t|_S = 0$, while the first model still shows that the conducting surface is located at distance d from the reference plane.

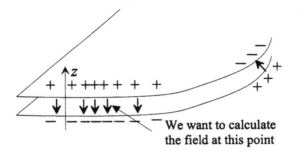

We want to calculate
the field at this point

Figure 2.3 The field inside a thin material layer at a given position is mainly dependent on charges and currents in a near vicinity. The fields of distant sources are mainly concentrated near those sources.

Many approximate results in this chapter will be based on the fact that the field distribution across various thin layers can be assumed to be *locally quasistatic*. Let us explain the meaning of this assumption here. Consider a layer whose thickness is much smaller compared to the wavelength and to the curvature of the layer (Figure 2.3). Note that the dimensions in the direction orthogonal to the interfaces can be arbitrarily large, so that wave phenomena cannot be neglected. Suppose we know the charge and current distribution on the two surfaces of the layer, and we want to find the fields inside the layer. The main idea here is that the field at a given point is mainly determined by the charges and currents just above and below the point of interest. Contributions from distant sources nearly cancel out (see Figure 2.3). This means that for the calculation of the field distribution across the layer thickness we can approximately replace the actual charges and currents by uniform distributions. Furthermore, if the thickness is small compared to the wavelength, we can assume that the distribution across the layer thickness is governed by the quasistatic equations. We will call this model *locally quasistatic*, because the quasistatic law is valid only locally: Waves are allowed to travel along the layer, but at every point the field distribution over the thickness remains the same. It is important to observe that these two assumptions are exactly the same as those adopted in the conventional transmission-line theory. In that theory, the wave equation defines waves along one direction (the axis of the transmission line), and in the other two directions, orthogonal to the line axis, the distribution is locally quasistatic. In modeling layers, we have the quasistatic law only for one direction across the slab.

Instead of starting our analysis from charges and currents on the surface, we can start from known tangential components of electric and magnetic fields on the two bounding surfaces of a layer (as is obvious from Huygens' principle). The distribution across the layer is easy to find in the quasistatic approximation. Indeed, the field components tangential to the layer boundaries (we will see later that there is no need to calculate the distribution of the normal components) obey the Helmholtz equation

$$\nabla^2 \mathbf{E}_t + k^2 \mathbf{E}_t = 0 \tag{2.1}$$

where index t denotes the component tangential to the interfaces. Let us assume for simplicity that the layer material is an isotropic medium, then $k^2 = \omega^2 \epsilon \mu$. The material parameters ϵ and μ can be complex numbers to account for losses or conductivity. The assumption of locality means that at every position we only consider dependence on just one direction, which is orthogonal to the layer (let it be direction z). The fields along the other two axes are taken as uniform. Thus, (2.1) simplifies as

$$\frac{\partial^2}{\partial z^2} \mathbf{E}_t + k^2 \mathbf{E}_t = 0 \tag{2.2}$$

Since we have assumed the quasistatic law for this distribution, we can neglect the second term of (2.2). For lossy or conductive layers this also means that the skin depth is assumed to be much larger than the layer thickness.[1] Hence, we get

$$\frac{\partial^2}{\partial z^2} \mathbf{E}_t = 0 \tag{2.3}$$

From (2.3) we conclude that $\mathbf{E}_t$ inside the layer varies approximately linearly over z: a very natural result for a thin dielectric layer. Based on this approximation, impedance boundary conditions for a layer can be built. This will be done in the next section.

2.2 SECOND-ORDER IMPEDANCE BOUNDARY CONDITIONS FOR THIN MATERIAL LAYERS

Here we introduce approximate second-order impedance boundary conditions that simulate thin dielectric layers. Consider a thin isotropic dielectric layer with scalar parameters ϵ and μ (Figure 2.4). The layer thickness is assumed to be small compared to the wavelength inside the slab's material. Let us find out how the tangential fields on the opposite sides of the layer are related, using the locally quasistatic approximation. Such relations, which connect *tangential* field components, are called *impedance* boundary or transition conditions.

---

[1] These limitations will be relaxed later in this chapter.

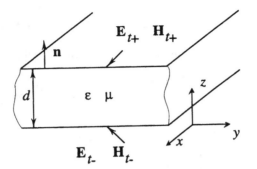

Figure 2.4 Geometry of a thin dielectric layer and the main notations.

2.2.1 Vector Transmission-Line Equations

Some general preliminary derivations are needed before we can establish boundary conditions for thin slabs. These results will also be rather important for building other models in the following chapters.

Starting from the frequency-domain Maxwell equations for the fields inside the slab

$$\nabla \times \mathbf{E} = -j\omega\mu\mathbf{H}, \qquad \nabla \times \mathbf{H} = j\omega\epsilon\mathbf{E} \qquad (2.4)$$

we separate the tangential $(\mathbf{E}_t, \mathbf{H}_t)$ and the normal (E_n, H_n) field components. This is needed because we want to obtain impedance boundary conditions, which connect tangential field components on the interfaces. The normal components should be eliminated. Thus, let us write

$$\mathbf{E} = \mathbf{E}_t + \mathbf{n}E_n, \quad \mathbf{H} = \mathbf{H}_t + \mathbf{n}H_n \qquad (2.5)$$

where

$$\mathbf{E}_t \cdot \mathbf{n} = 0, \qquad \mathbf{H}_t \cdot \mathbf{n} = 0 \qquad (2.6)$$

and also split the nabla operator:

$$\nabla = \nabla_t + \frac{\partial}{\partial z}\mathbf{n}, \qquad \nabla_t = \frac{\partial}{\partial x}\mathbf{x}_0 + \frac{\partial}{\partial y}\mathbf{y}_0 \qquad (2.7)$$

Equating separately the normal components of (2.4), we can express the normal components in terms of the tangential ones:

$$\mathbf{n}H_n = -\frac{1}{j\omega\mu}\nabla_t \times \mathbf{E}_t, \qquad \mathbf{n}E_n = \frac{1}{j\omega\epsilon}\nabla_t \times \mathbf{H}_t \qquad (2.8)$$

Next, we eliminate the normal components substituting (2.8) into the tangential part of (2.4). The result is

$$\frac{\partial}{\partial z}\mathbf{n} \times \mathbf{E}_t = -j\omega\mu\mathbf{H}_t - \frac{1}{j\omega\epsilon}\nabla_t \times (\nabla_t \times \mathbf{H}_t) \qquad (2.9)$$

$$\frac{\partial}{\partial z}\mathbf{n} \times \mathbf{H}_t = j\omega\epsilon\mathbf{E}_t + \frac{1}{j\omega\mu}\nabla_t \times (\nabla_t \times \mathbf{E}_t) \tag{2.10}$$

Cross-multiplying the first equation by $-\mathbf{n}$, this transforms to

$$\frac{\partial}{\partial z}\mathbf{E}_t = \left(j\omega\mu\overline{\overline{I}}_t + \frac{j}{\omega\epsilon}\nabla_t\nabla_t\right) \cdot (\mathbf{n} \times \mathbf{H}_t)$$

$$\frac{\partial}{\partial z}\mathbf{n} \times \mathbf{H}_t = \left(j\omega\epsilon\overline{\overline{I}}_t + \frac{j}{\omega\mu}\mathbf{n} \times \nabla_t\,\mathbf{n} \times \nabla_t\right) \cdot \mathbf{E}_t \tag{2.11}$$

These two equations can be called *vector transmission-line equations* because they connect "vector voltage" $\mathbf{E}_t$ and "vector current" $\mathbf{n} \times \mathbf{H}_t$ in the same way as the usual voltage and current are related in the conventional transmission-line equations. For plane-wave solutions the dependence on the transverse coordinates is exponential: $\exp(-j\mathbf{k}_t \cdot \mathbf{r})$, and the differential operator ∇_t is replaced by multiplication by $-j\mathbf{k}_t$:

$$\frac{\partial}{\partial z}\mathbf{E}_t = \left(j\omega\mu\overline{\overline{I}}_t - \frac{j}{\omega\epsilon}\mathbf{k}_t\mathbf{k}_t\right) \cdot (\mathbf{n} \times \mathbf{H}_t)$$

$$\frac{\partial}{\partial z}\mathbf{n} \times \mathbf{H}_t = \left(j\omega\epsilon\overline{\overline{I}}_t - \frac{j}{\omega\mu}\mathbf{n} \times \mathbf{k}_t\,\mathbf{n} \times \mathbf{k}_t\right) \cdot \mathbf{E}_t \tag{2.12}$$

Here $\overline{\overline{I}}_t = \overline{\overline{I}} - \mathbf{nn}$ is the two-dimensional unit dyadic.

The second-order wave equation for the transverse electric field component $\mathbf{E}_t$ immediately follows from the transmission-line equations (2.12) after elimination of $\mathbf{n} \times \mathbf{H}_t$, and it takes the form

$$\frac{\partial^2}{\partial z^2}\mathbf{E}_t + \beta^2\mathbf{E}_t = 0 \tag{2.13}$$

with

$$\beta^2 = \omega^2\epsilon\mu - k_t^2 \tag{2.14}$$

Now we see that the z-component of the propagation constant of eigenwaves equals $\beta = \pm\sqrt{\omega^2\epsilon\mu - k_t^2}$. Next, we substitute these values of the longitudinal propagation factors in the vector transmission-line equations (2.12), and find that the transverse fields in these eigenwaves depend on each other through wave impedances or admittances:

$$\mathbf{E}_t = \mp\overline{\overline{Z}} \cdot \mathbf{n} \times \mathbf{H}_t, \qquad \mathbf{n} \times \mathbf{H}_t = \mp\overline{\overline{Y}} \cdot \mathbf{E}_t \tag{2.15}$$

where the dyadic impedances and admittances are diagonal:

$$\overline{\overline{Z}} = Z^{TM}\frac{\mathbf{k}_t\mathbf{k}_t}{k_t^2} + Z^{TE}\frac{\mathbf{n} \times \mathbf{k}_t\,\mathbf{n} \times \mathbf{k}_t}{k_t^2}, \qquad \overline{\overline{Y}} = \overline{\overline{Z}}^{-1} \tag{2.16}$$

and the upper and lower signs correspond to the waves propagating in the positive and the negative directions of the z-axis, correspondingly. The characteristic impedances for the two eigenpolarizations read

$$Z^{TM} = \sqrt{\frac{\mu}{\epsilon}} \sqrt{1 - \frac{k_t^2}{\omega^2 \epsilon \mu}}, \qquad Z^{TE} = \sqrt{\frac{\mu}{\epsilon}} \frac{1}{\sqrt{1 - \frac{k_t^2}{\omega^2 \epsilon \mu}}} \qquad (2.17)$$

In many applications we deal with uniaxial slabs, such that their permittivities and permeabilities are different for the fields along the axis defined by $\mathbf{n}$ (ϵ_n, μ_n) and in the transverse plane (ϵ_t, μ_t). For this case the derivation is similar, and the results read

$$\frac{\partial}{\partial z} \mathbf{E}_t = \left(j\omega \mu_t \overline{\overline{I}}_t + \frac{j}{\omega \epsilon_n} \nabla_t \nabla_t \right) \cdot (\mathbf{n} \times \mathbf{H}_t)$$

$$\frac{\partial}{\partial z} \mathbf{n} \times \mathbf{H}_t = \left(j\omega \epsilon_t \overline{\overline{I}}_t + \frac{j}{\omega \mu_n} \mathbf{n} \times \nabla_t \mathbf{n} \times \nabla_t \right) \cdot \mathbf{E}_t \qquad (2.18)$$

The wave equation is

$$\frac{\partial^2}{\partial z^2} \mathbf{E}_t + \left(\beta_{TM}^2 \frac{\mathbf{k}_t \mathbf{k}_t}{k_t^2} + \beta_{TE}^2 \frac{\mathbf{n} \times \mathbf{k}_t \, \mathbf{n} \times \mathbf{k}_t}{k_t^2} \right) \cdot \mathbf{E}_t = 0 \qquad (2.19)$$

Because the dyadic in the last equation is diagonal, the eigensolutions of (2.13) are obviously two linearly polarized vectors: one is parallel to $\mathbf{k}_t$ and another one to $\mathbf{n} \times \mathbf{k}_t$, just like in the case of an isotropic medium. The first solution corresponds to a TM-polarized wave, with the magnetic field orthogonal to the $\mathbf{k}_t$ vector, and the second one is a TE-wave (see Section 2.4). β_{TM} and β_{TE} are the normal components of the propagation factors for the TM- and TE-polarized eigenwaves, respectively:

$$\beta_{TM}^2 = \frac{\epsilon_t}{\epsilon_n} \left(\omega^2 \epsilon_n \mu_t - k_t^2 \right), \qquad \beta_{TE}^2 = \frac{\mu_t}{\mu_n} \left(\omega^2 \epsilon_t \mu_n - k_t^2 \right) \qquad (2.20)$$

The difference as compared to the isotropic medium is that these two values are different for the two eigenpolarizations. The wave impedances are

$$Z^{TM} = \sqrt{\frac{\mu_t}{\epsilon_t}} \sqrt{1 - \frac{k_t^2}{\omega^2 \epsilon_n \mu_t}} \qquad (2.21)$$

$$Z^{TE} = \sqrt{\frac{\mu_t}{\epsilon_t}} \frac{1}{\sqrt{1 - \frac{k_t^2}{\omega^2 \epsilon_t \mu_n}}} \qquad (2.22)$$

2.2.2   Derivation of Impedance Conditions

The next important step is the *averaging* of the tangential field components across the slab thickness. Let us simply integrate the equations for the tangential field components (2.9) and (2.10) and define the averaged fields as

$$\widehat{\mathbf{E}}_t = \frac{1}{d} \int\limits_0^d \mathbf{E}_t \, dz, \qquad \widehat{\mathbf{H}}_t = \frac{1}{d} \int\limits_0^d \mathbf{H}_t \, dz \qquad (2.23)$$

After averaging, we get

$$\frac{\mathbf{n} \times \mathbf{E}_{t+} - \mathbf{n} \times \mathbf{E}_{t-}}{d} = -j\omega\mu\widehat{\mathbf{H}}_t - \frac{1}{j\omega\epsilon}\nabla_t \times (\nabla_t \times \widehat{\mathbf{H}}_t) \qquad (2.24)$$

$$\frac{\mathbf{n} \times \mathbf{H}_{t+} - \mathbf{n} \times \mathbf{H}_{t-}}{d} = j\omega\epsilon\widehat{\mathbf{E}}_t + \frac{1}{j\omega\mu}\nabla_t \times (\nabla_t \times \widehat{\mathbf{E}}_t) \qquad (2.25)$$

Now we need to connect the averaged fields and the fields on the interfaces in order to find boundary conditions. Under the assumption of locally quasistatic field behavior, the equation for the tangential fields is simply $\partial^2 \mathbf{E}_t/\partial z^2 = 0$ (2.3), which means the linear variation of the field across the slab thickness. This gives

$$\widehat{\mathbf{E}}_t = \frac{\mathbf{E}_{t+} + \mathbf{E}_{t-}}{2}, \qquad \widehat{\mathbf{H}}_t = \frac{\mathbf{H}_{t+} + \mathbf{H}_{t-}}{2} \qquad (2.26)$$

Consider for simplicity a slab on an ideally conducting surface, then the tangential electric field component $\mathbf{E}_{t-} = 0$. Equation (2.24) shows that $\mathbf{E}_{t+}$ is a small quantity of the order $O(kd)$, where $k = \omega\sqrt{\epsilon\mu}$ (tangential electric field near a metal plane is small). From (2.25) we see that the difference $\mathbf{H}_{t+} - \mathbf{H}_{t-} = O(kd)^2$ is of the second order of smallness; thus, this difference should be neglected: $\widehat{\mathbf{H}}_t \approx \mathbf{H}_{t+}$. With this in mind, from (2.24) we finally get the impedance relation between tangential fields on the free interface of the slab:

$$\mathbf{n} \times \mathbf{E}_{t+} = -j\omega d\mu\mathbf{H}_{t+} - \frac{d}{j\omega\epsilon}\nabla_t \times (\nabla_t \times \mathbf{H}_{t+}) \qquad (2.27)$$

This result can be written in a simpler-looking form:

$$\mathbf{E}_{t+} = \overline{\overline{Z}}_s \cdot \mathbf{n} \times \mathbf{H}_{t+} = \left(j\omega\mu d\,\overline{\overline{I}}_t - \frac{d}{j\omega\epsilon}\nabla_t\nabla_t \right) \cdot \mathbf{n} \times \mathbf{H}_{t+} \qquad (2.28)$$

In the Cartesian coordinate system this result reads

$$E_{x+} = -j\omega d\mu H_{y+} + \frac{d}{j\omega\epsilon}\left(\frac{\partial^2}{\partial x^2}H_{y+} - \frac{\partial^2}{\partial x\partial y}H_{x+} \right) \qquad (2.29)$$

$$E_{y+} = j\omega d\mu H_{x+} + \frac{d}{j\omega\epsilon}\left(\frac{\partial^2}{\partial y\partial x}H_{y+} - \frac{\partial^2}{\partial y^2}H_{x+} \right) \qquad (2.30)$$

For the more general case when both surfaces can be open we get, from (2.24) and (2.25),

$$\mathbf{E}_{t+} - \mathbf{E}_{t-} = \left(j\omega\mu d \overline{\overline{I}}_t - \frac{d}{j\omega\epsilon} \nabla_t \nabla_t \right) \cdot \mathbf{n} \times \widehat{\mathbf{H}}_t \qquad (2.31)$$

$$\mathbf{n} \times \mathbf{H}_{t+} - \mathbf{n} \times \mathbf{H}_{t-} = j\omega\epsilon d \widehat{\mathbf{E}}_t + \frac{d}{j\omega\mu} \nabla_t \times (\nabla_t \times \widehat{\mathbf{E}}_t) \qquad (2.32)$$

We could of course simply replace the averaged tangential fields here by their approximate values from (2.26). That would result in boundary conditions containing only the fields on the interfaces. Note, however, that the right-hand side of (2.31) is of the first order of smallness (kd is the small parameter). From the second equation in this set we see that the difference between the tangential magnetic fields on the two sides is also of the first order of smallness. In view of (2.26), we conclude that

$$\widehat{\mathbf{H}}_t = \mathbf{H}_{t-} + O(kd) \qquad (2.33)$$

This correction, after substitution in (2.31), will give a second-order correction to the tangential electric field values. Because the second-order terms have been neglected in deriving (2.26), this correction should also be neglected here. The same conclusion concerns (2.32). Finally, we have

$$\mathbf{E}_{t+} - \mathbf{E}_{t-} = \left(j\omega\mu d \overline{\overline{I}}_t - \frac{d}{j\omega\epsilon} \nabla_t \nabla_t \right) \cdot \mathbf{n} \times \mathbf{H}_{t-} \qquad (2.34)$$

$$\mathbf{n} \times \mathbf{H}_{t+} - \mathbf{n} \times \mathbf{H}_{t-} = j\omega\epsilon d \mathbf{E}_{t-} + \frac{d}{j\omega\mu} \nabla_t \times (\nabla_t \times \mathbf{E}_{t-}) \qquad (2.35)$$

Fields in the right-hand side can be replaced by $\mathbf{E}_{t+}$ and $\mathbf{H}_{t+}$ without a loss of accuracy.

The second-order operator impedance boundary conditions can be used in analytical investigations of various systems with thin layers or implemented numerically (e.g., within finite-difference time-domain simulators), as we will see in Chapter 3.

2.2.3 Other Forms of Approximate Boundary Conditions for Thin Layers

Looking at the right-hand side of equations (2.29) and (2.30) we can recognize terms that are equal to the normal electric field component, because

$$E_z = \frac{1}{j\omega\epsilon} \left(\frac{\partial H_y}{\partial x} - \frac{\partial H_x}{\partial y} \right) \qquad (2.36)$$

This observation allows us to rewrite the boundary conditions as

$$E_{x+} = -j\omega d\mu H_{y+} + d\left.\frac{\partial E_z}{\partial x}\right|_{z=d-} \tag{2.37}$$

$$E_{y+} = j\omega d\mu H_{x+} + d\left.\frac{\partial E_z}{\partial y}\right|_{z=d-} \tag{2.38}$$

It is important to observe that the boundary conditions (2.29) and (2.30) have been derived considering the fields *inside* the slab. Since the tangential fields are continuous on the interface, in the impedance boundary conditions we can substitute the fields just on the interface but *outside* the material layer. However, it is not so in the case of (2.37) and (2.38). To obtain the boundary conditions for the fields outside the layer we have to apply the boundary condition for the normal electric field component. Assuming that the region $z > d$ is filled by air, so that $\epsilon E_z|_{z=d-} = \epsilon_0 E_z|_{z=d+}$ or, in our notations, $\epsilon E_z|_{z=d-} = \epsilon_0 E_{z+}$, we can finally write the boundary conditions as

$$E_{x+} = -j\omega d\mu H_{y+} + d\frac{\epsilon_0}{\epsilon}\frac{\partial E_{z+}}{\partial x} \tag{2.39}$$

$$E_{y+} = j\omega d\mu H_{x+} + d\frac{\epsilon_0}{\epsilon}\frac{\partial E_{z+}}{\partial y} \tag{2.40}$$

These boundary conditions were introduced by L.A. Vainshtein (Weinstein) [3, Chapter 9].

A disadvantage of this form is that the conditions depend on materials *outside* the layer. In contrast, the impedance boundary conditions (2.29) and (2.30) depend only on the layer properties and can be directly applied when the materials outside the layers are arbitrary. This provides a tool to analyze situations when a layer is located in complex environments such as anisotropic or even nonlinear media.

A note on the terminology is appropriate here. Deriving conditions (2.29) and (2.30) in the beginning of this section, we called them *second-order* boundary conditions because they contained second-order spatial derivatives of the fields on the interface. However, we have just seen that the same result can be expressed in terms of the first derivatives, if the normal field component is involved. To avoid any ambiguity, we will use this kind of classification only for impedance boundary conditions.

Vainshtein derived conditions (2.39) and (2.40) in a different way, which is very instructive, and we will need that approach for our goals here. To derive approximate boundary conditions for a thin layer on a conducting ground plane, we can start from the boundary condition on the ideally conducting surface

$$\mathbf{E}_t|_{z=0} = 0, \qquad \text{or} \quad E_{x,y}|_{z=0} = 0 \tag{2.41}$$

Inside the slab, the fields should satisfy the Maxwell equations that allow us to find the derivatives of the tangential field components with respect to the vertical coordinate z. Assuming that the layer is filled by an isotropic dielectric and using the Maxwell equation

$$\frac{\partial E_z}{\partial y} - \frac{\partial E_y}{\partial z} = -j\omega\mu H_x \tag{2.42}$$

we find

$$\frac{\partial E_y}{\partial z} = j\omega\mu H_x + \frac{\partial E_z}{\partial y} \tag{2.43}$$

Similarly,

$$\frac{\partial E_x}{\partial z} = -j\omega\mu H_y + \frac{\partial E_z}{\partial x} \tag{2.44}$$

Now, assuming that the thickness of the layer is small, we use the first term of the Taylor expansion to find the fields on the upper surface:

$$E_{x,y}|_{z=0} = E_{x,y}|_{z=d-} - \frac{\partial E_{x,y}}{\partial z}\bigg|_{z=d-} d \tag{2.45}$$

Using (2.43) and (2.44), we arrive at (2.37) and (2.38), and finally to the approximate boundary conditions (2.39) and (2.40).

With the help of the same trick we can write effective boundary conditions on the surface $z = 0$, instead of the upper free surface $z = d$, as has been done by Vainshtein for conditions (2.39) and (2.40). To write the boundary conditions at $z = 0$ we should move the reference plane from $z = d$ to $z = 0$. To find how the fields change we can again make use of the Taylor expansion (2.45), but for fields in free space (more generally, in the medium filling the space at $z > d$). This can be understood from Figure 2.2: In the boundary condition model formulated at $z = 0$, there is free space in $0 < z < d$. With this in mind, we assume that at $z = d$ the boundary conditions (2.39) and (2.40) are satisfied, and apply (2.45) to find the *equivalent* fields at $z = 0$. The result is

$$E_{x+} = -j\omega d(\mu - \mu_0)H_{y+} + d\left(\frac{\epsilon_0}{\epsilon} - 1\right)\frac{\partial E_{z+}}{\partial x} \tag{2.46}$$

$$E_{y+} = j\omega d(\mu - \mu_0)H_{x+} + d\left(\frac{\epsilon_0}{\epsilon} - 1\right)\frac{\partial E_{z+}}{\partial y} \tag{2.47}$$

Clearly, if the material parameters of the layer are the same as that of the surrounding medium, so there is actually no layer at all, the last conditions reduce to the conditions on an ideally conducting surface. If the conditions are formulated at $z = d$, the result is nontrivial, which reflects the fact that the reference plane is at distance d from the ground plane, where the tangential electric field is not zero.

In a similar way we can reformulate the impedance boundary conditions (2.29) and (2.30), so that they hold at $z = 0$. Making use of the Taylor expansion (2.45), rewritten as

$$E_{x,y}|_{z=d} = E_{x,y}|_{z=0} + \frac{\partial E_{x,y}}{\partial z}\bigg|_{z=0} d \qquad (2.48)$$

we substitute the derivatives from (2.43) and (2.44), and then eliminate the normal field component E_z using (2.36). The result is

$$E_x = -j\omega d(\mu - \mu_0)H_y + \frac{d}{j\omega}\left(\frac{\epsilon_0}{\epsilon} - 1\right)\left(\frac{\partial^2}{\partial x^2}H_y - \frac{\partial^2}{\partial x \partial y}H_x\right) \qquad (2.49)$$

$$E_y = j\omega d(\mu - \mu_0)H_x + \frac{d}{j\omega}\left(\frac{\epsilon_0}{\epsilon} - 1\right)\left(\frac{\partial^2}{\partial y \partial x}H_y - \frac{\partial^2}{\partial y^2}H_x\right) \qquad (2.50)$$

In vector form,

$$\mathbf{E}_t = \left[j\omega d(\mu - \mu_0)\overline{\overline{I}}_t - \frac{d}{j\omega}\left(\frac{\epsilon_0}{\epsilon} - 1\right)\nabla_t\nabla_t\right] \cdot \mathbf{n} \times \mathbf{H}_t \qquad (2.51)$$

Let us stress again that these conditions are valid only if there is free space at $z > d$.

In the same way, we can reformulate the boundary conditions (2.34) and (2.35) for a layer in free space, so that all fields are measured on the lower interface:

$$\mathbf{E}_{t+} = \mathbf{E}_{t-} + \left[j\omega(\mu - \mu_0)d\overline{\overline{I}}_t - \frac{d}{j\omega}\left(\frac{\epsilon_0}{\epsilon} - 1\right)\nabla_t\nabla_t\right] \cdot \mathbf{n} \times \mathbf{H}_{t-} \qquad (2.52)$$

$$\mathbf{n} \times \mathbf{H}_{t+} = \mathbf{n} \times \mathbf{H}_{t-} + j\omega(\epsilon - \epsilon_0)d\mathbf{E}_{t-} + \frac{d}{j\omega}\left(\frac{\mu_0}{\mu} - 1\right)\nabla_t \times (\nabla_t \times \mathbf{E}_{t-}) \qquad (2.53)$$

2.2.4   Example: Partially Filled Waveguide

Here we present a simple example of the use of impedance boundary conditions in a study of eigenwaves of a partially filled rectangular waveguide with a small height (Figure 2.5).

To solve for the eigenwaves, we apply the boundary conditions (2.29) and (2.30) twice, to model both the dielectric layer and the air gap. In this application, it is obviously an advantage to apply conditions formulated at the open surface of the layers, and not at the waveguide walls. This is because the two layers have a common interface, and it is enough to demand the tangential electric fields to be continuous on the interface. Looking for the general solution in the form $\exp(-jk_x x - jk_y y)$, the continuity requirement for E_x and E_y takes the form

$$-j\omega d_2 \mu H_{y+} + \frac{d_2}{j\omega\epsilon}(-k_x^2 H_{y+} + k_x k_y H_{x+})$$

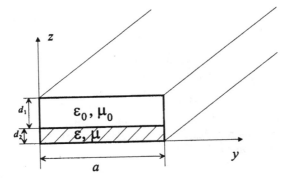

Figure 2.5 Partially filled waveguide of a small height $(d_1 + d_2 \ll \lambda)$ with ideally conducting walls.

$$= j\omega d_1\mu_0 H_{y+} - \frac{d_1}{j\omega\epsilon_0}(-k_x^2 H_{y+} + k_x k_y H_{x+}) \qquad (2.54)$$

$$j\omega d_2\mu H_{x+} + \frac{d_2}{j\omega\epsilon}(k_y^2 H_{x+} - k_x k_y H_{y+})$$

$$= -j\omega d_1\mu_0 H_{x+} - \frac{d_1}{j\omega\epsilon_0}(k_y^2 H_{x+} - k_x k_y H_{y+}) \qquad (2.55)$$

This is a uniform linear system of equations for two unknowns: H_{x+} and H_{y+}. Equating the determinant of the system to zero, we get a relation between k_x and k_y:

$$k_x^2 + k_y^2 = \omega^2 \frac{d_2\mu + d_1\mu_0}{\frac{d_2}{\epsilon} + \frac{d_1}{\epsilon_0}} \qquad (2.56)$$

From the form of this equation we see that the two solutions for k_y differ only by sign. Thus, all the field components depend on the y-coordinate as

$$E_x = Ae^{-jk_y y} + Be^{jk_y y} \qquad (2.57)$$

where A and B are constants, and k_y is a solution to (2.56). On the side walls of the waveguide, at $y = 0$ and $y = a$, tangential to the wall electric field E_x must be zero:

$$A + B = 0, \qquad Ae^{-jk_y a} + Be^{jk_y a} = 0 \qquad (2.58)$$

Equating the determinant of this system to zero, we find that

$$k_y = \frac{n\pi}{a}, \qquad n = \pm 1, \pm 2, \pm 3, \ldots \qquad (2.59)$$

The solution $n = 0$ is excluded because in that case the field is identically zero.

Finally, we substitute (2.59) in (2.56) to find the propagation constants of the eigenmodes:

$$k_x = \sqrt{\omega^2 \frac{d_2\mu + d_1\mu_0}{\frac{d_2}{\epsilon} + \frac{d_1}{\epsilon_0}} - \left(\frac{n\pi}{a}\right)^2}, \qquad n = 1, 2, 3, \ldots \qquad (2.60)$$

This result has a clear and simple physical meaning: The solution is determined by averaged permittivities and permeabilities. Otherwise, the solution has the same form as for simple rectangular waveguides. Only modes with almost linear variation of the fields across the small vertical size of the waveguide (or almost uniform field in that direction) can be found by this method. For waveguides with small heights considered here, these are the modes of practical interest.

The different averaging rules for the permittivity and permeability in (2.60) can be understood in terms of the line capacitance and inductance per unit length. Consider a planar waveguide with the same layered filling but no side walls. Formally this corresponds to $a \to \infty$ in the above theory. In this waveguide the fundamental mode is a quasi transverse electromagnetic (quasi-TEM) wave, and we can introduce transmission-line parameters, so that $k_x = \omega\sqrt{LC}$, where L and C are the inductance and capacitance per unit length. The capacitance per unit length (and unit width) C can be calculated in a trivial way as the capacitance between two conducting planes with a two-layer filling. Let the charge density on the planes be $\pm q$, then the normal components of the electric fields in the two slabs are, from the boundary condition on metal, $E_z = q/\epsilon_0$ and $E_z = q/\epsilon$. The voltage between the plates is, obviously, $q(d_2/\epsilon + d_1/\epsilon_0)$. From here

$$C = \frac{1}{\frac{d_2}{\epsilon} + \frac{d_1}{\epsilon_0}} \qquad (2.61)$$

In the same way, because the total magnetic flux through the unit section of the line is the sum of the fluxes through the two layers, we find that

$$L = d_2\mu + d_1\mu_0 \qquad (2.62)$$

This corresponds to the series connection of two inductances and the parallel connection of two capacitances corresponding to the two layers.

In the theory of the waveguide with a finite width a, the waves are allowed to travel along y, and of course no quasi-TEM solution can exist. However, the field distribution over the small height remains *locally* quasistatic. That is why in (2.60) we can recognize the same expression as follows from the quasistatic analysis.

2.2.5   More Complicated Layers

The locally quasistatic approximation together with the field averaging allows us to find simple approximate boundary conditions for rather complicated layers. The power of this approach is mainly due to the fact that the averaged fields and the fields on the two sides of a thin layer can be related in a simple way even for layers filled by complex materials or for inhomogeneous layers.

Ferrite Layer

As a first example, let us consider a thin planar layer filled by a magnetized ferrite material. For a saturated ferrite medium the permeability is a dyadic

$$\overline{\overline{\mu}} = \mu_0 \left[\mu \overline{\overline{I}} - (\mu - 1) \mathbf{h}_0 \mathbf{h}_0 - j \mu_a \mathbf{h}_0 \times \overline{\overline{I}} \right] \tag{2.63}$$

where $\overline{\overline{I}}$ is the unit dyadic, $\mathbf{h}_0$ is the unit vector parallel to the constant bias magnetic field, and the relative parameters μ and μ_a are

$$\mu = \frac{\omega_H (\omega_H + \omega_M) - \omega^2}{\omega_H^2 - \omega^2}, \qquad \mu_a = \frac{\omega_M \omega}{\omega_H^2 - \omega^2} \tag{2.64}$$

Two constants ω_H and ω_M are proportional to the constant magnetic field strength H_0 and the saturation magnetization M_0, respectively: $\omega_H = \gamma H_0$, $\omega_M = \gamma M_0$, with γ being the gyromagnetic ratio. In the matrix representation, if the direction of the constant magnetic field is written in the spherical coordinate system by the angles θ and ϕ, the components of the $[\mu]$ matrix read

$$\mu_{xx} = \mu_0 [\mu + (1 - \mu) \sin^2 \theta \cos^2 \phi] \tag{2.65}$$

$$\mu_{yy} = \mu_0 [\mu + (1 - \mu) \sin^2 \theta \sin^2 \phi] \tag{2.66}$$

$$\mu_{zz} = \mu_0 [\mu + (1 - \mu) \cos^2 \theta] \tag{2.67}$$

$$\mu_{xy} = \mu_{yx}^* = \mu_0 [(1 - \mu) \sin^2 \theta \sin \phi \cos \phi + j \mu_a \cos \theta] \tag{2.68}$$

$$\mu_{xz} = \mu_{zx}^* = \mu_0 [(1 - \mu) \sin \theta \cos \theta \cos \phi - j \mu_a \sin \theta \sin \phi] \tag{2.69}$$

$$\mu_{yz} = \mu_{zy}^* = \mu_0 [(1 - \mu) \sin \theta \cos \theta \sin \phi + j \mu_a \sin \theta \cos \phi] \tag{2.70}$$

where * denotes the complex conjugate value.

The Helmholtz equation for the fields in the ferrite layer is of course different from that for a simple dielectric:

$$\nabla^2 \mathbf{E} + \omega^2 \epsilon \overline{\overline{\mu}} \cdot \mathbf{E} = 0, \qquad \nabla^2 \mathbf{H} + \omega^2 \epsilon \overline{\overline{\mu}} \cdot \mathbf{H} = 0 \tag{2.71}$$

but in the locally *quasistatic* approximation the term containing the permeability dyadic is neglected[2], so the relation between the fields on the boundaries and the averaged field is still the same as for the simple isotropic case:

$$\widehat{\mathbf{E}}_t = \frac{\mathbf{E}_{t+} + \mathbf{E}_{t-}}{2}, \qquad \widehat{\mathbf{H}}_t = \frac{\mathbf{H}_{t+} + \mathbf{H}_{t-}}{2} \tag{2.72}$$

---

[2]This means that the approximation is not valid near the ferromagnetic resonance where components of the permeability dyadic can take large values, and the wavelength inside the slab can become very small.

The boundary conditions can now be derived from the averaged Maxwell equations in the same way as for a simple dielectric slab. For a ferrite slab magnetized in the direction normal to the interfaces ($\mathbf{h}_0 = \mathbf{n}$), we get [4]:

$$\mathbf{E}_{t+} = \mathbf{E}_{t-} + j\omega\mu_0\mu d \left(\overline{\overline{I}}_t - j\frac{\mu_a}{\mu}\mathbf{n} \times \overline{\overline{I}}_t + \frac{1}{\omega^2\epsilon\mu_0\mu}\nabla_t\nabla_t \right) \cdot \mathbf{n} \times \mathbf{H}_{t-} \quad (2.73)$$

$$\mathbf{n} \times \mathbf{H}_{t+} = \mathbf{n} \times \mathbf{H}_{t-} + j\omega\epsilon d \left(\overline{\overline{I}}_t + \frac{1}{\omega^2\epsilon\mu_0}\mathbf{n} \times \nabla_t \mathbf{n} \times \nabla_t \right) \cdot \mathbf{E}_{t-} \quad (2.74)$$

For a thin slab positioned on an ideally conducting surface we have, neglecting the second-order terms,

$$\mathbf{E}_{t+} = j\omega\mu_0\mu d \left(\overline{\overline{I}}_t - j\frac{\mu_a}{\mu}\mathbf{n} \times \overline{\overline{I}}_t + \frac{1}{\omega^2\epsilon\mu_0\mu}\nabla_t\nabla_t \right) \cdot \mathbf{n} \times \mathbf{H}_{t+} \quad (2.75)$$

The averaging method can be applied to structures with nonplanar geometries as well; for example, a coaxial cable with ferrite filling is analyzed in [5]. Layers of other complex media can be modeled in a similar way [6, 7].

Inhomogeneous Dielectric Layer

Suppose that a thin layer has an inhomogeneous filling, so that its parameters depend on the coordinate z, normal to the two interfaces:

$$\epsilon = \epsilon(z), \qquad \mu = \mu(z) \qquad (2.76)$$

To extend the previous results for this case we only need to find a relation between the fields on the slab boundaries and the averaged fields inside the layer. In the locally quasistatic approximation, we assume again that the fields inside depend mainly on the charges and currents just on the nearest area of the opposite interfaces of the slab. In this approximation we can treat the transversal fields as functions of coordinate z only. Because the layer is thin, we can use the quasistatic theory and introduce a scalar potential ϕ: $\mathbf{E}(z) = -\nabla\phi(x, y, z)$, where the dependence of ϕ on the transverse coordinates x and y is linear. The equation for the potential

$$\nabla \cdot [\epsilon(z)\nabla\phi(x, y, z)] = 0 \qquad (2.77)$$

can be easily solved by separating the variables. Since the dependence on x and y is linear, we get

$$\frac{d}{dz}\left[\epsilon(z)\frac{d\phi(z)}{dz} \right] = 0 \qquad (2.78)$$

From here,

$$\frac{d\phi(z)}{dz} = \frac{C_1}{\epsilon(z)}, \qquad \phi(x, y, z) = \left[\int_0^z \frac{C_1}{\epsilon(z)}\,dz + C_2 \right] xy \qquad (2.79)$$

and the tangential field inside the slab reads

$$\mathbf{E}_t(z) = \mathbf{C}_1 \int_0^z \frac{dz}{\epsilon(z)} + \mathbf{C}_2 \tag{2.80}$$

where $\mathbf{C}_{1,2}$ are constant tangential vectors. For the normal component, we have from (2.79)

$$E_n(z) = \frac{C}{\epsilon(z)} \tag{2.81}$$

where C is a constant scalar.

Equation (2.81) tells us that $D_n = \epsilon(z)E_n = C = \text{const}(z)$. Thus, the averaging is trivial in this case:

$$\widehat{D}_n = \check{\epsilon}\widehat{E}_n \tag{2.82}$$

with

$$\check{\epsilon} = \frac{d}{\int_0^d \frac{dz}{\epsilon(z)}} \tag{2.83}$$

Constant vectors $\mathbf{C}_{1,2}$ in (2.80) can be found assuming the tangential fields on the boundaries are known:

$$\mathbf{E}_t(z) = \mathbf{E}_{t-} + (\mathbf{E}_{t+} - \mathbf{E}_{t-}) \left[\int_0^d \frac{dz}{\epsilon(z)} \right]^{-1} \int_0^z \frac{dz}{\epsilon(z)} \tag{2.84}$$

Next we integrate this equation to find the averaged field, which leads to

$$\widehat{\mathbf{E}}_t = g(\epsilon)\,\mathbf{E}_{t+} + [1 - g(\epsilon)]\,\mathbf{E}_{t-} \tag{2.85}$$

where $g(\epsilon)$ is a functional of $\epsilon(z)$ defined as

$$g(\epsilon) = \frac{\check{\epsilon}}{d^2} \int_0^d \int_0^t \frac{dz}{\epsilon(z)}\, dt \tag{2.86}$$

In the same way, we find after integration the averaged tangential displacement vector:

$$\widehat{\mathbf{D}}_t = l(\epsilon)\,\mathbf{E}_{t+} + m(\epsilon)\,\mathbf{E}_{t-} \tag{2.87}$$

with

$$l(\epsilon) = \frac{\check{\epsilon}}{d^2} \int_0^d \epsilon(z) \int_0^z \frac{dt}{\epsilon(t)}\, dz, \qquad m(\epsilon) = \frac{1}{d} \int_0^d \epsilon(z)\, dz - l(\epsilon) \tag{2.88}$$

Note that if $\epsilon(z) = \text{const}(z)$, we have $g(\epsilon) = 1/2$, which coincides with the previous result for uniform layers (2.26).

The same results can be written for the magnetic field by changing $\mathbf{E}$ to $\mathbf{H}$, $\mathbf{D}$ to $\mathbf{B}$, and introducing

$$\breve{\mu} = \frac{d}{\int_0^d \frac{dz}{\mu(z)}} \tag{2.89}$$

and

$$g(\mu) = \frac{\breve{\mu}}{d^2} \int_0^d \int_0^t \frac{dz}{\mu(z)} \, dt \tag{2.90}$$

Now we are ready to derive the approximate boundary conditions. First, we write the Maxwell equations for the normal and tangential components, as was done for the homogeneous slab in (2.8)–(2.10):

$$\mathbf{n} B_n = -\frac{1}{j\omega} \nabla_t \times \mathbf{E}_t, \qquad \mathbf{n} D_n = \frac{1}{j\omega} \nabla_t \times \mathbf{H}_t \tag{2.91}$$

$$\frac{\partial}{\partial z} \mathbf{n} \times \mathbf{E}_t = -j\omega \mathbf{B}_t - \nabla_t \times \mathbf{n} E_n \tag{2.92}$$

$$\frac{\partial}{\partial z} \mathbf{n} \times \mathbf{H}_t = j\omega \mathbf{D}_t - \nabla_t \times \mathbf{n} H_n \tag{2.93}$$

Integrating from 0 to d we find for the averaged fields

$$\mathbf{n} \widehat{B}_n = -\frac{1}{j\omega} \nabla_t \times \widehat{\mathbf{E}}_t, \qquad \mathbf{n} \widehat{D}_n = \frac{1}{j\omega} \nabla_t \times \widehat{\mathbf{H}}_t \tag{2.94}$$

$$\frac{1}{d}(\mathbf{n} \times \mathbf{E}_{t+} - \mathbf{n} \times \mathbf{E}_{t-}) = -j\omega \widehat{\mathbf{B}}_t - \nabla_t \times \mathbf{n} \widehat{E}_n \tag{2.95}$$

$$\frac{1}{d}(\mathbf{n} \times \mathbf{H}_{t+} - \mathbf{n} \times \mathbf{H}_{t-}) = j\omega \widehat{\mathbf{D}}_t - \nabla_t \times \mathbf{n} \widehat{H}_n \tag{2.96}$$

The next step is to use (2.82) that connects the averaged normal field components and the averaged displacements to eliminate the normal components. Finally, we can repeat the procedure applied above in the case of a homogeneous slab with the new averaging rule for the tangential field components (2.85)–(2.87). The resulting boundary conditions [11] read

$$\mathbf{E}_{t+} = \mathbf{E}_{t-} + j\omega \widehat{\mu} d \left(\overline{\overline{I}}_t + \frac{1}{\omega^2 \breve{\epsilon} \breve{\mu}} \nabla_t \nabla_t \right) \cdot \mathbf{n} \times \mathbf{H}_{t-} \tag{2.97}$$

$$\mathbf{n} \times \mathbf{H}_{t+} = \mathbf{n} \times \mathbf{H}_{t-} + j\omega \widehat{\epsilon} d \left[\mathbf{E}_{t-} + \frac{1}{\omega^2 \breve{\epsilon} \breve{\mu}} \nabla_t \times (\nabla_t \times \mathbf{E}_{t-}) \right] \tag{2.98}$$

where

$$\widehat{\epsilon} = l(\epsilon) + m(\epsilon) = \frac{1}{d} \int_0^d \epsilon(z) \, dz, \qquad \widehat{\mu} = l(\mu) + m(\mu) = \frac{1}{d} \int_0^d \mu(z) \, dz \tag{2.99}$$

are the averaged parameters of the slab. For a layer on an ideally conducting surface we have

$$\mathbf{E}_{t+} = j\omega\widehat{\mu}d\left(\overline{\overline{I}}_t + \frac{1}{\omega^2\breve{\epsilon}\breve{\mu}}\nabla_t\nabla_t\right)\cdot\mathbf{n}\times\mathbf{H}_{t+} \qquad (2.100)$$

An alternative approach to the modeling of nonhomogeneous layers can be found in [8]. More accurate models can be developed for multilayer coverings using the exact field solution for planar dielectric layers [9, 10].

2.3 EXACT BOUNDARY CONDITIONS

For plane waves,[3] boundary conditions for planar dielectric layers can be derived without any approximations. This can be done following basically the same steps as in the derivation of the second-order impedance boundary conditions in the locally quasistatic approximation. Indeed, the only approximation assumed in that approach was in the relation between the averaged fields and the fields on the interfaces. Let us find the corresponding exact relation.

2.3.1 Exact "Averaging"

If the slab is excited by a plane wave with the tangential component of the wave vector $\mathbf{k}_t$, waves in the slab will have the same tangential component of the wave vector. For an isotropic slab, the normal to the interfaces component of the wave vector inside the slab is $\beta = \sqrt{k^2 - k_t^2}$ ($k = \omega\sqrt{\epsilon\mu}$). The general solution for the tangential electric field inside the slab is

$$\mathbf{E}_t(z) = \mathbf{A}e^{-j\beta z} + \mathbf{B}e^{j\beta z} \qquad (2.101)$$

(we are only interested in the z-dependence now). Constant vectors $\mathbf{A}$ and $\mathbf{B}$ are determined by the boundary conditions

$$\mathbf{E}(0) = \mathbf{E}_{t-}, \qquad \mathbf{E}(d) = \mathbf{E}_{t+} \qquad (2.102)$$

which lead to

$$\mathbf{A} = \frac{\mathbf{E}_{t+} - \mathbf{E}_{t-}e^{j\beta d}}{e^{-j\beta d} - e^{j\beta d}}, \qquad \mathbf{B} = -\frac{\mathbf{E}_{t+} - \mathbf{E}_{t-}e^{-j\beta d}}{e^{-j\beta d} - e^{j\beta d}} \qquad (2.103)$$

Next, we simply integrate (2.101) over z from 0 to d to find the averaged field. The result reads:

$$\widehat{\mathbf{E}}_t = \frac{1}{d}\left[\mathbf{A}\frac{1}{-j\beta}\left(e^{-j\beta d} - 1\right) + \mathbf{B}\frac{1}{j\beta}\left(e^{j\beta d} - 1\right)\right]$$

---

[3]In other words, in the spatial Fourier domain.

$$= (\mathbf{E}_{t+} + \mathbf{E}_{t-}) \frac{2 - e^{j\beta d} - e^{-j\beta d}}{j\beta d\,(e^{-j\beta d} - e^{j\beta d})} = (\mathbf{E}_{t+} + \mathbf{E}_{t-}) \frac{\tan \frac{\beta d}{2}}{\beta d} \qquad (2.104)$$

For $|\beta|d \ll 1$ we use the Taylor expansion

$$\frac{\tan \frac{\beta d}{2}}{\beta d} = \frac{1}{2} + \frac{1}{24}(\beta d)^2 + \dots \qquad (2.105)$$

and in the first-order approximation recover the simple relation derived in the locally quasistatic approximation. In the previous theory the second-order and higher-order terms have been neglected. That is why we *had to* neglect all the other second-order terms in that theory. Also, this result shows that the small parameter in the the locally quasistatic boundary conditions for thin layers is $|\beta|d$, which means that for the validity of the second-order formulas based on the locally quasistatic approximation, the slab thickness must be small compared to the wavelength in the *normal* direction inside the slab.

2.3.2   Boundary Conditions

The exact boundary condition follows from (2.24) and (2.25) if we use the exact relation (2.104) for the averaged fields. For slabs on ideally conducting surfaces we write

$$\frac{\mathbf{n} \times \mathbf{E}_{t+}}{d} = -j\omega\mu(\mathbf{H}_{t+} + \mathbf{H}_{t-})f(\beta d) + \frac{1}{j\omega\epsilon}\mathbf{k}_t \times [\mathbf{k}_t \times (\mathbf{H}_{t+} + \mathbf{H}_{t-})]\,f(\beta d)$$

$$\qquad (2.106)$$

$$\frac{\mathbf{n} \times \mathbf{H}_{t+} - \mathbf{n} \times \mathbf{H}_{t-}}{d} = j\omega\epsilon\mathbf{E}_{t+}f(\beta d) - \frac{1}{j\omega\mu}\mathbf{k}_t \times (\mathbf{k}_t \times \mathbf{E}_{t+})f(\beta d) \quad (2.107)$$

where we have denoted

$$f(\beta d) = \frac{\tan \frac{\beta d}{2}}{\beta d} \qquad (2.108)$$

Now we can eliminate $\mathbf{H}_{t-}$ and find the relation between the Fourier transformed tangential fields $\mathbf{E}_{t+}$ and $\mathbf{H}_{t+}$ on the free interface. After some algebra needed to eliminate $\mathbf{H}_{t-}$ (more details in [11]), the result is

$$\mathbf{E}_{t+} = j\omega\mu\frac{\tan\beta d}{\beta}\left(\overline{\overline{I}}_t - \frac{\mathbf{k}_t\mathbf{k}_t}{k^2}\right)\cdot\mathbf{n} \times \mathbf{H}_{t+} \qquad (2.109)$$

This is the exact boundary condition but only for Fourier transformed fields (that is, for plane waves). One cannot simply transform it to the physical space replacing $\mathbf{k}_t \to -j\nabla_t$, because $\beta = \sqrt{k^2 - k_t^2}$ and we end up with a pseudodifferential operator:

$$\mathbf{E}_{t+} = j\omega\mu\frac{\tan(\sqrt{k^2 + \nabla_t^2}\,d)}{\sqrt{k^2 + \nabla_t^2}}\left(\overline{\overline{I}}_t + \frac{\nabla_t\nabla_t}{k^2}\right)\cdot\mathbf{n} \times \mathbf{H}_{t+} \qquad (2.110)$$

The only way to use it in practice is to formally expand the impedance in the Taylor series or use another rational approximation with respect to ∇_t. Clearly, the exact expression will contain tangential field derivatives of all orders, up to infinity. This is an expected result, because the fields in the physical system which we model interact globally: In principle, the field at any chosen position depends on the sources *everywhere*. We try to model this by a *local* impedance relation that contains the field values at a single point only. As a result, we naturally arrive to an operator impedance: To calculate the result, the knowledge of the global field behavior is required anyway.

However, having the exact relation (2.109) we can introduce appropriate rational approximations of $f(\beta d)$ tailored for specific applications. Such approximations make physical and practical sense because we know that in thin layers the locality assumption is indeed valid. Before discussing such models (in the next chapter), we will introduce a vector circuit description for a slab. Conceptually this description is similar to the vector transmission-line model where the transverse electric and magnetic fields are considered as vector voltages and currents of an equivalent transmission line with dyadic parameters. For a half-space the vector transmission-line model was introduced by I. Lindell and E. Alanen [12] and used in later developments of the *exact image method* [13, Chapter 7]. Here we will use vector voltages and currents in an equivalent *circuit* model of material layers.

2.4 VECTOR CIRCUIT INTERPRETATION: LAYERS BETWEEN ARBITRARY MEDIA

The exact plane-wave solution for a planar slab (2.106) and (2.107) is a linear relation between tangential fields on the two sides of the slab. We can write it in matrix form:

$$\left(\begin{array}{c} \mathbf{E}_{t+} \\ \mathbf{n} \times \mathbf{H}_{t+} \end{array} \right) = \left(\begin{array}{cc} \bar{\bar{a}}_{11} & \bar{\bar{a}}_{12} \\ \bar{\bar{a}}_{21} & \bar{\bar{a}}_{22} \end{array} \right) \cdot \left(\begin{array}{c} \mathbf{E}_{t-} \\ \mathbf{n} \times \mathbf{H}_{t-} \end{array} \right) \tag{2.111}$$

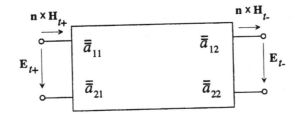

Figure 2.6 Transmission-matrix model of a slab.

Interpreting $\mathbf{E}_{t-}$ as a *vector* input voltage, $\mathbf{n} \times \mathbf{H}_{t-}$ as a *vector* input current, and the fields on the opposite side as the output voltage and current, we can draw an equivalent circuit (Figure 2.6), where the properties of the layer are described by a transmission matrix with the dyadic components

$$\overline{\overline{a}}_{11} = \overline{\overline{a}}_{22} = \cos(\beta d)\overline{\overline{I}}_t \qquad (2.112)$$

$$\overline{\overline{a}}_{12} = j\frac{k\eta}{\beta}\sin(\beta d)\overline{\overline{A}}, \qquad \overline{\overline{a}}_{21} = j\frac{k}{\eta\beta}\sin(\beta d)\overline{\overline{C}} \qquad (2.113)$$

where $\eta = \sqrt{\mu/\epsilon}$ is the wave impedance in the material of the slab, and dyadics

$$\overline{\overline{A}} = \overline{\overline{I}}_t - \frac{\mathbf{k}_t\mathbf{k}_t}{k^2} = \frac{\beta^2}{k^2}\frac{\mathbf{k}_t\mathbf{k}_t}{k_t^2} + \frac{\mathbf{n}\times\mathbf{k}_t\,\mathbf{n}\times\mathbf{k}_t}{k_t^2} \qquad (2.114)$$

$$\overline{\overline{C}} = \overline{\overline{I}}_t - \frac{\mathbf{n}\times\mathbf{k}_t\,\mathbf{n}\times\mathbf{k}_t}{k^2} = \frac{\mathbf{k}_t\mathbf{k}_t}{k_t^2} + \frac{\beta^2}{k^2}\frac{\mathbf{n}\times\mathbf{k}_t\,\mathbf{n}\times\mathbf{k}_t}{k_t^2} \qquad (2.115)$$

Here we have written the two-dimensional unit dyadic in the basis formed by two orthogonal unit vectors $\mathbf{k}_t/|k_t|$ and $\mathbf{n}\times\mathbf{k}_t/|k_t|$:

$$\overline{\overline{I}}_t = \frac{\mathbf{k}_t\,\mathbf{k}_t}{k_t^2} + \frac{\mathbf{n}\times\mathbf{k}_t\,\mathbf{n}\times\mathbf{k}_t}{k_t^2} \qquad (2.116)$$

This is the most convenient basis in this case because it is formed by the vectors naturally defined by the geometry ($\mathbf{n}$) and the incident wave ($\mathbf{k}_t$).[4] Furthermore, the scalar coefficients in this basis have a clear physical meaning: They correspond to the two main orthogonal polarizations of incident plane waves. This can be seen from Figures 2.7 and 2.8. In case of the TM (or *parallel*) polarization of the incident wave, transverse field vectors $\mathbf{E}_t$ and $\mathbf{n}\times\mathbf{H}_t$ are parallel to the transverse component of the wave vector $\mathbf{k}_t$. For the other (TE or *perpendicular*) polarization, these vectors are parallel to the other basis vector, $\mathbf{n}\times\mathbf{k}_t$.

We observe that all the dyadic coefficients $\overline{\overline{a}}_{ij}$ share a common set of eigenvectors, which are $\mathbf{k}_t/|k_t|$ and $\mathbf{n}\times\mathbf{k}_t/|k_t|$. In the basis of the eigenvectors these dyadics become diagonal. The corresponding eigennumbers are the slab parameters for TM and TE polarized incident fields, respectively. Dyadics $\overline{\overline{a}}_{12}$ and $\overline{\overline{a}}_{21}$ (2.113) can be rewritten in the diagonal form:

$$\overline{\overline{a}}_{12} = j\frac{k\eta}{\beta}\sin(\beta d)\overline{\overline{A}} = a_{12}^{TM}\frac{\mathbf{k}_t\mathbf{k}_t}{k_t^2} + a_{12}^{TE}\frac{\mathbf{n}\times\mathbf{k}_t\,\mathbf{n}\times\mathbf{k}_t}{k_t^2} \qquad (2.117)$$

where

$$a_{12}^{TM} = j\eta\frac{\beta}{k}\sin(\beta d), \qquad a_{12}^{TE} = j\eta\frac{k}{\beta}\sin(\beta d) \qquad (2.118)$$

---

[4]Dyadics $\overline{\overline{a}}_{11}$ and $\overline{\overline{a}}_{22}$ are proportional to the transverse unit dyadic $\overline{\overline{I}}_t$, so they can be treated as scalars.

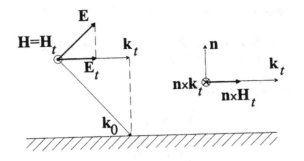

Figure 2.7 TM-polarized plane wave incident on a planar interface. Vectors $\mathbf{k}_t$ and $\mathbf{n} \times \mathbf{k}_t$ form a two-dimensional basis. Both $\mathbf{E}_t$ and $\mathbf{n} \times \mathbf{H}_t$ are parallel to $\mathbf{k}_t$.

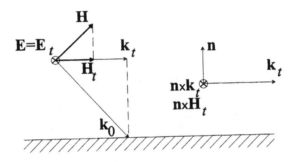

Figure 2.8 TE-polarized plane wave incident on a planar interface. In this case $\mathbf{E}_t$ and $\mathbf{n} \times \mathbf{H}_t$ are parallel to $\mathbf{n} \times \mathbf{k}_t$.

are the scalar coefficients for TM and TE polarized waves, respectively. For $\overline{\overline{a}}_{21}$ we find

$$\overline{\overline{a}}_{21} = a_{21}^{TM} \frac{\mathbf{k}_t \mathbf{k}_t}{k_t^2} + a_{21}^{TE} \frac{\mathbf{n} \times \mathbf{k}_t\, \mathbf{n} \times \mathbf{k}_t}{k_t^2} \tag{2.119}$$

where

$$a_{21}^{TM} = j \frac{k}{\eta \beta} \sin(\beta d), \qquad a_{21}^{TE} = j \frac{\beta}{\eta k} \sin(\beta d) \tag{2.120}$$

Note that the inverse transmission matrix can be found simply by changing the sign of the unit vector $\mathbf{n}$:

$$\begin{pmatrix} \mathbf{E}_{t-} \\ -\mathbf{n} \times \mathbf{H}_{t-} \end{pmatrix} = \begin{pmatrix} \overline{\overline{a}}_{11} & \overline{\overline{a}}_{12} \\ \overline{\overline{a}}_{21} & \overline{\overline{a}}_{22} \end{pmatrix} \cdot \begin{pmatrix} \mathbf{E}_{t+} \\ -\mathbf{n} \times \mathbf{H}_{t+} \end{pmatrix} \tag{2.121}$$

We can also express electric fields in terms of magnetic fields, solving linear equations (2.111) for electric fields. This leads to

$$\begin{pmatrix} \mathbf{E}_{t+} \\ \mathbf{E}_{t-} \end{pmatrix} = \begin{pmatrix} \overline{\overline{Z}}_{11} & \overline{\overline{Z}}_{12} \\ \overline{\overline{Z}}_{21} & \overline{\overline{Z}}_{22} \end{pmatrix} \cdot \begin{pmatrix} \mathbf{n} \times \mathbf{H}_{t+} \\ \mathbf{n} \times \mathbf{H}_{t-} \end{pmatrix} \tag{2.122}$$

where the elements of the impedance matrix read

$$\overline{\overline{Z}}_{11} = \overline{\overline{a}}_{11} \cdot \overline{\overline{a}}_{21}^{-1} \tag{2.123}$$

$$\overline{\overline{Z}}_{12} = -\overline{\overline{a}}_{11} \cdot \overline{\overline{a}}_{21}^{-1} \cdot \overline{\overline{a}}_{22} + \overline{\overline{a}}_{12} \tag{2.124}$$

$$\overline{\overline{Z}}_{21} = \overline{\overline{a}}_{21}^{-1} \tag{2.125}$$

$$\overline{\overline{Z}}_{22} = -\overline{\overline{a}}_{21}^{-1} \cdot \overline{\overline{a}}_{22} \tag{2.126}$$

Substitution of $\overline{\overline{a}}_{ij}$ gives

$$\overline{\overline{Z}}_{11} = -\overline{\overline{Z}}_{22} = -j\frac{\eta k}{\beta} \cot(\beta d)\,\overline{\overline{A}} \tag{2.127}$$

$$\overline{\overline{Z}}_{12} = -\overline{\overline{Z}}_{21} = j\frac{\eta k}{\beta} \frac{1}{\sin(\beta d)}\,\overline{\overline{A}} \tag{2.128}$$

$$\overline{\overline{Z}}_{11} + \overline{\overline{Z}}_{12} = -\overline{\overline{Z}}_{22} - \overline{\overline{Z}}_{21} = j\frac{\eta k}{\beta} \tan\left(\frac{\beta d}{2}\right)\overline{\overline{A}} \tag{2.129}$$

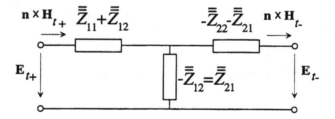

Figure 2.9 T-circuit model of a slab.

The impedance matrix can be visualized in terms of its equivalent T-circuit (see Figure 2.9). In the same way, admittance matrix can be introduced. Solving for the magnetic fields in terms of electric ones, we get

$$\begin{pmatrix} \mathbf{n} \times \mathbf{H}_{t+} \\ \mathbf{n} \times \mathbf{H}_{t-} \end{pmatrix} = \begin{pmatrix} \overline{\overline{Y}}_{11} & \overline{\overline{Y}}_{12} \\ \overline{\overline{Y}}_{21} & \overline{\overline{Y}}_{22} \end{pmatrix} \cdot \begin{pmatrix} \mathbf{E}_{t+} \\ \mathbf{E}_{t-} \end{pmatrix} \tag{2.130}$$

where the elements of the impedance matrix read

$$\overline{\overline{Y}}_{11} = \overline{\overline{a}}_{22} \cdot \overline{\overline{a}}_{12}^{-1} \tag{2.131}$$

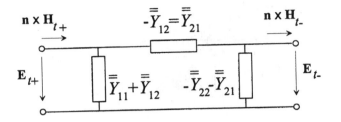

Figure 2.10 Π-circuit model of a slab.

$$\overline{\overline{Y}}_{12} = \overline{\overline{a}}_{21} - a_{22} \cdot \overline{\overline{a}}_{12}^{-1} \cdot \overline{\overline{a}}_{11} \tag{2.132}$$

$$\overline{\overline{Y}}_{21} = \overline{\overline{a}}_{12}^{-1} \tag{2.133}$$

$$\overline{\overline{Y}}_{22} = -\overline{\overline{a}}_{12}^{-1} \cdot \overline{\overline{a}}_{11} \tag{2.134}$$

Finally,

$$\overline{\overline{Y}}_{11} = -\overline{\overline{Y}}_{22} = j\frac{k}{\eta\beta}\cot(\beta d)\,\overline{\overline{C}} \tag{2.135}$$

$$\overline{\overline{Y}}_{12} = -\overline{\overline{Y}}_{21} = -j\frac{k}{\eta\beta}\frac{1}{\sin(\beta d)}\,\overline{\overline{C}} \tag{2.136}$$

$$\overline{\overline{Y}}_{11} + \overline{\overline{Y}}_{12} = -\overline{\overline{Y}}_{22} - \overline{\overline{Y}}_{21} = j\frac{k}{\eta\beta}\tan\left(\frac{\beta d}{2}\right)\overline{\overline{C}} \tag{2.137}$$

The equivalent Π-circuit is shown in Figure 2.10.

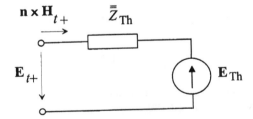

Figure 2.11 Thévenin equivalent voltage source: another vector circuit representation for an isotropic slab.

Rearranging (2.111), we can also write the following linear relations between the fields on the two sides of the slab:

$$\mathbf{E}_{t+} = \frac{1}{\cos(\beta d)}\mathbf{E}_{t-} + j\frac{\eta k}{\beta}\tan(\beta d)\,\overline{\overline{A}}\cdot(\mathbf{n}\times\mathbf{H}_{t+}) \tag{2.138}$$

$$\mathbf{n}\times\mathbf{H}_{t+} = \frac{1}{\cos(\beta d)}\mathbf{n}\times\mathbf{H}_{t-} + j\frac{k}{\eta\beta}\tan(\beta d)\,\overline{\overline{C}}\cdot\mathbf{E}_{t+} \tag{2.139}$$

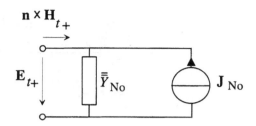

Figure 2.12 Norton equivalent current source.

These relations have a clear vector circuit interpretation. The first one corresponds to a voltage source (Figure 2.11). Here, the source voltage is

$$\mathbf{E}_{\text{Th}} = \frac{1}{\cos(\beta d)} \mathbf{E}_{t-} \qquad (2.140)$$

and the source impedance reads

$$\overline{\overline{Z}}_{\text{Th}} = j\frac{\eta k}{\beta} \tan(\beta d)\, \overline{\overline{A}} \qquad (2.141)$$

Formula (2.139) corresponds to a Norton equivalent circuit, Figure 2.12, with

$$\mathbf{J}_{\text{No}} = -\frac{1}{\cos(\beta d)} \mathbf{n} \times \mathbf{H}_{t-}, \qquad \overline{\overline{Y}}_{\text{No}} = j\frac{k}{\eta\beta} \tan(\beta d)\, \overline{\overline{C}} \qquad (2.142)$$

From this form we can conveniently derive boundary conditions on slabs backed by ideal electric and magnetic walls. If $\mathbf{E}_{t-} = 0$, (2.138) gives the exact boundary condition on the upper side of the slab:

$$\mathbf{E}_{t+} = j\frac{\eta k}{\beta} \tan(\beta d)\, \overline{\overline{A}} \cdot (\mathbf{n} \times \mathbf{H}_{t+}) \qquad (2.143)$$

which is the same as (2.109). For a slab on an ideal magnetic conductor $(\mathbf{n} \times \mathbf{H}_{t-} = 0)$, we get from (2.139)

$$\mathbf{n} \times \mathbf{H}_{t+} = j\frac{k}{\eta\beta} \tan(\beta d)\, \overline{\overline{C}} \cdot \mathbf{E}_{t+} \qquad (2.144)$$

2.5 THIN SHEETS OF DIELECTRICS, MAGNETICS, AND METALS

Thin sheet models can be applied when we can neglect the slab thickness and consider the layer as an infinitely thin sheet. This implies that either permittivity or permeability or both are very large. Sheet models can be found from the exact solution for plane waves.

Let us assume that $\mu = \mu_0$, but $\epsilon = \epsilon_0 \epsilon_r$ and $|\epsilon_r|$ is very large. This means that $|k|$ is very large as compared to $|k_t|$. Indeed, $\mathbf{k}_t$ is the tangential component of the wave vector in the surrounding medium (which we assume to be free space or some medium with not too large $|\epsilon_r|$ and $|\mu_r|$). This means that $\beta = \sqrt{k^2 - k_t^2} \approx k$, and $\overline{\overline{A}} \approx \overline{\overline{C}} \approx \overline{\overline{I}}_t$. Both facts are extremely pleasant since (1) all the impedances are now scalar quantities; and (2) the sheet conditions can be trivially transformed to the physical space (no differential operators at all).

Taking the appropriate limit in (2.127)–(2.129), we find

$$Z_{11} = -j\eta \cot(kd), \qquad Z_{12} = j\eta \frac{1}{\sin(kd)}, \qquad Z_{11} + Z_{12} = j\eta \tan\left(\frac{kd}{2}\right)$$

$$(2.145)$$

Finally, suppose that although $|\epsilon_r|$ is large, $|k|d \ll 1$ because d is very small. Then

$$Z_{11} + Z_{12} \approx 0 \qquad (2.146)$$

and

$$Z_{12} = j\eta \frac{1}{kd} = j\eta_0 \frac{1}{\epsilon_r k_0 d} \qquad (2.147)$$

is large in the absolute value. Here $\eta_0 = \sqrt{\mu_0/\epsilon_0}$ and $k_0 = \omega\sqrt{\epsilon_0\mu_0}$ are the free-space parameters.

Under this assumption, the electric fields on the two sides of the slab are nearly the same, but the tangential magnetic fields are discontinuous:

$$\mathbf{E}_{t+} \approx \mathbf{E}_{t-}, \qquad \mathbf{n}\times\mathbf{H}_{t+}-\mathbf{n}\times\mathbf{H}_{t-} \approx j\frac{\sin(kd)}{\eta}\mathbf{E}_{t-} \approx \frac{j}{\eta}kd\,\mathbf{E}_{t-} = \frac{j}{\eta_0}\epsilon_r k_0 d\,\mathbf{E}_{t-}$$

$$(2.148)$$

We have derived this result approximating the exact relations for a black-box model (Figure 2.1). This means that the fields marked by $t+$ are measured at $z = d$, and those marked by $t-$ are defined at the other side of the slab, at $z = 0$. As here we model *sheets*, it is more natural to use the sheet model, writing conditions for the fields at a certain equivalent infinitely thin surface. This can be done using the Taylor expansion for the fields in free space, similar to what was done in Section 2.2.3. Let us choose the plane of one of the layer sides as such an equivalent surface. To be specific, let us write the equivalent conditions on the lower surface. Considering one of the tangential components of the magnetic field, we write the Taylor expansion keeping the first term only:

$$H_{x+} = H_{x-} + \left.\frac{\partial H_x}{\partial z}\right|_{z=+0} d \qquad (2.149)$$

From the Maxwell equation,

$$\frac{\partial H_x}{\partial z} \approx j\omega\epsilon_0 E_y = \frac{j}{\eta_0}k_0 E_y \qquad (2.150)$$

Thus,

$$\mathbf{n} \times \mathbf{H}_{t+}|_{z=d} = \mathbf{n} \times \mathbf{H}_{t+}|_{z=+0} + \frac{j}{\eta_0} k_0 \mathbf{E}_{t+}|_{z=+0} \qquad (2.151)$$

Finally, substituting into (2.148), we arrive at the sheet conditions where the fields are measured on the two sides of an equivalent infinitely thin surface located at $z = 0$:

$$\mathbf{E}_{t+} \approx \mathbf{E}_{t-}, \qquad \mathbf{n}\times\mathbf{H}_{t+} - \mathbf{n}\times\mathbf{H}_{t-} \approx d\left(\frac{j}{\eta}k - \frac{j}{\eta_0}k_0\right)\mathbf{E}_{t-} = \frac{j}{\eta_0}(\epsilon_r - 1)k_0 d\,\mathbf{E}_{t-}$$

$$(2.152)$$

In the last expression it has been assumed that the slab is nonmagnetic: $\mu = \mu_0$. For simplicity of notations, we keep the same notation $\mathbf{H}_{t+}$ as in the black-box model (2.148), although in (2.148) the magnetic field is measured at $z = d$, but in (2.152) it is defined at $z = +0$.

The equivalent T-circuit for a dielectric sheet is shown in Figure 2.13.

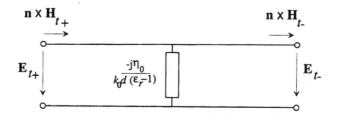

Figure 2.13 Equivalent shunt impedance for a dielectric sheet.

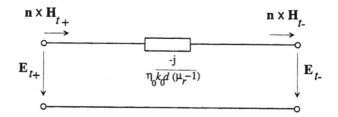

Figure 2.14 Equivalent series admittance for a magnetic sheet.

For highly conductive material sheets, substituting $\epsilon_r - 1 \approx -j\sigma/(\omega\epsilon_0)$, where σ is the conductivity, we have

$$\mathbf{n} \times \mathbf{H}_{t+} - \mathbf{n} \times \mathbf{H}_{t-} \approx \sigma d\,\mathbf{E}_{t-} \qquad (2.153)$$

This is simply the Ohm law for a unit-width strip of the sheet. The left-hand side expression equals the current flowing along this strip, σd is the resistance of the strip (per unit length), and $\mathbf{E}_{t-}$ is the voltage per unit length.

Magnetic sheets (large $|\mu_r|$) or combined sheets (both $|\epsilon_r|$ and $|\mu_r|$ are large) can be studied in the same way. For a magnetic sheet, we get

$$\mathbf{H}_{t+} \approx \mathbf{H}_{t-}, \qquad \mathbf{E}_{t+} - \mathbf{E}_{t-} \approx jk_0 d\eta_0 (\mu_r - 1)\mathbf{n} \times \mathbf{H}_{t-} \qquad (2.154)$$

The corresponding equivalent circuit is shown in Figure 2.14.

PROBLEMS

2.1 Check that (2.10) and (2.11) are equivalent.

2.2 Derive (2.28) from (2.27).

2.3 Derive approximate boundary conditions for a thin uniaxial dielectric layer:

$$\overline{\overline{\epsilon}} = \epsilon_1 \overline{\overline{I}}_t + \epsilon_2 \mathbf{a}_0 \mathbf{a}_0, \qquad \mu = \mu_0$$

where $\mathbf{a}_0$ is a unit vector, $\overline{\overline{I}}_t = \overline{\overline{I}} - \mathbf{a}_0 \mathbf{a}_0$, and $\overline{\overline{I}}$ is the three-dimensional unit dyadic. Consider the case when vector $\mathbf{a}_0$ is orthogonal to the slab boundaries. The case when $\mathbf{a}_0$ is in the plane of the slab is slightly more complicated.

2.4 Find the exact relation connecting tangential electric and magnetic fields on the free surface of a uniaxial slab (as in the previous problem) backed by an ideally conducting boundary.

2.5 Prove (2.123)–(2.126) and (2.127)–(2.129).

2.6 Prove (2.121).

2.7 Derive an exact relation for the propagation factors of a planar waveguide with ideally conducting walls and a two-layer isotropic filling. Hint: Use the exact boundary conditions for a slab on an ideally conducting plane.

References

[1] Senior, T.B.A., and J.L. Volakis, *Approximate Boundary Conditions in Electromagnetics*, London: The Institute of Electrical Engineers, 1995.

[2] Hoppe, D.J., and Y. Rahmat-Samii, *Impedance Boundary Conditions in Electromagentics*, Washington, D.C.: Taylor & Francis, 1995.

[3] Weinstein, L.A., *The Theory of Diffraction and the Factorization Method*, Boulder, CA: The Golem Press, 1969 (translation from Russian by P. Beckmann).

[4] Tretyakov, S.A., A.S. Cherepanov, and M.I. Oksanen, "Averaging Method for Analysing Waveguides with Anisotropic Filling," *Radio Science,* Vol. 26, No. 2, 1991, pp. 523-528.

[5] Tretyakov, S.A., M.I. Oksanen, and A.S. Cherepanov, "New Ferrite-Filled Waveguiding Structures Analysed by the Averaging Method," *IEE Proceedings, Part H,* Vol. 139, No. 3, 1992, pp. 227-232.

[6] Tretyakov, S.A., "Thin Pseudochiral Layers: Approximate Boundary Conditions and Potential Applications," *Microwave and Optical Technology Lett.,* Vol. 6, No. 2, 1993, pp. 112-115.

[7] Haliullin, D.Y., and S.A. Tretyakov, "Reflection and Transmission Coefficients for Thin Bianisotropic Layers," *IEE Proc. Microwaves, Antennas and Propagation,* Vol. 145, No. 2, 1998, pp. 163-168.

[8] Ammati, H., and S. He, "Effective Impedance Boundary Conditions for an Inhomogeneous Thin Layer on a Curved Metallic Surface," *IEEE Trans. Antennas and Propagation,* Vol. 46, No. 5, 1998, pp. 710-715.

[9] Tretyakov, S.A., "Generalized Impedance Boundary Conditions for Isotropic Multilayers," *Microwave and Optical Technology Lett.,* Vol. 17, No. 4, 1998, pp. 262-265.

[10] Puska, P.P., S.A. Tretyakov, and A.H. Sihvola, "Approximate Impedance Boundary Conditions for Isotropic Multilayered Media," *IEE Proc. Microwaves, Antennas and Propagation,* Vol. 146, No. 2, 1999, pp. 163-166.

[11] Oksanen M.I., S.A. Tretyakov, and I.V. Lindell, "Vector Circuit Theory for Isotropic and Chiral Slabs," *J. Electromagnetic Waves and Applications,* Vol. 4, No. 7, 1990, pp. 613-643.

[12] Lindell, I.V., and E. Alanen, "Exact Image Theory for the Sommerfeld Half-Space Problem, Part III: General Formulation," *IEEE Trans. Antennas and Propagation,* Vol. 32, 1984, pp. 1027-1032.

[13] Lindell, I.V., *Methods for Electromagnetic Field Analysis,* Piscataway, NJ: IEEE Press, 1995.

Chapter 3

Interfaces and Higher-Order Boundary Conditions

This chapter is about impedance models of interfaces with material bodies and more accurate (*higher-order*) models of material layers. These models are especially valuable if one of the materials is highly conductive or if the dielectric contrast between the two media is very high. The other case is the opposite situation when two media are very similar. In the limiting case when there is actually no interface at all because the two media are the same, an impedance model is an absorbing boundary condition, useful in numerical techniques.

Impedance and sheet models for interfaces and layers have been studied in the literature for a long time. The simplest approximate boundary condition for modeling conducting bodies was proposed probably for the first time in the late 1930s. In a textbook on radio wave propagation [1], A.N. Shchukin published a boundary condition that connected tangential and normal electric field components at interfaces with conductive bodies. S.M. Rytov derived an equivalent relation between tangential electric and magnetic fields [2]. In the introduction to his paper, Rytov said that the problem was suggested by M.A. Leontovich. After Leontovich published (in 1948) an extensive analysis of the new boundary condition, with a reference to Shchukin, and a discussion on its validity limitations [3], it became common to refer to this simple impedance boundary condition as the *Leontovich* boundary condition. Since those early days, the simple impedance model was extended to thin layers, and various more accurate higher-order models were developed and applied in many problems of applied electromagnetics (see monographs [4,5]). Furthermore, higher-order approximate boundary conditions were introduced for more complex layers, like layers of chiral [6,7] and bianisotropic [8] media, as well as for multilayer slabs [9–11]. Our approach here will be based on approximations of the exact boundary or transition conditions in the Fourier domain, similar to the method used in [6, 12].

3.1 HIGHER-ORDER IMPEDANCE MODELS FOR INTERFACES WITH CONDUCTORS

Impedance boundary conditions are widely used to model interfaces with highly conducting bodies (e.g., metals at microwaves and the Earth at radio frequencies), and there are several ways to derive them. In the simplest case, the impedance is just a scalar quantity, but a more accurate modeling leads to impedance operators which contain spatial derivatives of the fields on the interface. Such boundary conditions are called *higher-order impedance boundary conditions*. We will discuss them starting by establishing the *exact* boundary conditions for plane waves.

3.1.1 Exact Surface Impedance Operator

The exact surface impedance operator can be derived in the same way as we used in Chapter 2 for slabs. Remember that the averaged field equations (2.24) and (2.25) in Chapter 2 are exact and can be applied for slabs of arbitrary thickness and with arbitrary parameters. Here we will apply them for a half-space; that is, for an infinitely thick slab.

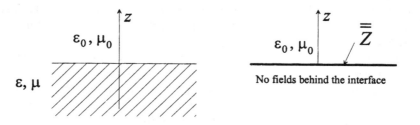

Figure 3.1 Half-space filled by an isotropic medium and its model as an impedance surface.

Consider a half-space of an isotropic medium (parameters ϵ, μ, $\eta = \sqrt{\mu/\epsilon}$, and $k = \omega\sqrt{\epsilon\mu}$, Figure 3.1). The z-axis is directed normally to the interface, and the interface is on the plane $z = 0$. Let us integrate (2.9) over z from minus infinity to zero, which is the same as the averaging applied before for a slab, but now the averaging region extends to infinity. The result is

$$\mathbf{n} \times \mathbf{E}_{t+} = -j\omega\mu \int_{-\infty}^{0} \mathbf{H}_t \, dz - \frac{1}{j\omega\epsilon} \nabla_t \times \left(\nabla_t \times \int_{-\infty}^{0} \mathbf{H}_t \, dz \right) \qquad (3.1)$$

Here, index $t+$ marks the field at the upper interface (which is, of course, the only interface in the present case). We assume that the material has some losses, so that the fields at $z = -\infty$ decay to zero. In the Fourier domain, for plane waves with the transverse component of the wave vector equal to $\mathbf{k}_t$,

$$\mathbf{n} \times \mathbf{E}_{t+} = -j\omega\mu \int_{-\infty}^{0} \mathbf{H}_t \, dz + \frac{1}{j\omega\epsilon} \mathbf{k}_t \times \left(\mathbf{k}_t \times \int_{-\infty}^{0} \mathbf{H}_t \, dz \right) \qquad (3.2)$$

The general solution for the fields in the half-space $z < 0$ is

$$\mathbf{H}_t = \mathbf{H}_{t+} e^{j\beta z}, \qquad \beta = \sqrt{k^2 - k_t^2}, \qquad \mathrm{Im}\{\beta\} < 0 \qquad (3.3)$$

Trivial integration gives

$$\int_{-\infty}^{0} \mathbf{H}_t \, dz = \mathbf{H}_{t+} \frac{1}{j\beta} \qquad (3.4)$$

Finally, substitution in (3.2) leads to the impedance boundary condition for plane waves:

$$\mathbf{E}_{t+} = \overline{\overline{Z}} \cdot \mathbf{n} \times \mathbf{H}_{t+} = \eta \frac{k}{\beta} \overline{\overline{A}} \cdot \mathbf{n} \times \mathbf{H}_{t+} = \eta \frac{k \left(\overline{\overline{I}}_t - \frac{\mathbf{k}_t \mathbf{k}_t}{k^2} \right)}{\sqrt{k^2 - k_t^2}} \cdot \mathbf{n} \times \mathbf{H}_{t+} \qquad (3.5)$$

Dyadic $\overline{\overline{A}}$ is defined by (2.114).

Formally transforming to the physical space, we replace $\mathbf{k}_t \Rightarrow -j\nabla_t$. The result is [6, 13]

$$\mathbf{E}_{t+} = \overline{\overline{Z}} \cdot \mathbf{n} \times \mathbf{H}_{t+} \qquad \text{with} \qquad \overline{\overline{Z}} = \eta \frac{k \left(\overline{\overline{I}}_t + \frac{\nabla_t \nabla_t}{k^2} \right)}{\sqrt{k^2 + \nabla_t^2}} \qquad (3.6)$$

This is the exact boundary condition that models the properties of the whole half-space in $z < 0$. In the exact model the impedance $\overline{\overline{Z}}$ is a pseudodifferential operator. This result should have been expected, because obviously the electric field at a certain point at an interface depends on the currents and charges on the whole plane. *Local* boundary conditions connecting electric and magnetic fields at a single point can model the interface properties only approximately. Operational boundary conditions which involve spatial derivatives can be more accurate, but the only way to write an exact boundary condition is to involve integrations over the interface or, alternatively, derivations in the interface plane with all orders. This is the case of the pseudodifferential operator (3.6). We already dealt with the same problem in Section 2.3.2.

The same result for an isotropic half-space can be obtained directly from the vector transmission-line model discussed in Chapter 2. The input impedance of a semi-infinite vector transmission line modeling a half-space equals its wave impedance. Indeed, notice that the surface impedance operator (3.5) simply equals the dyadic wave impedance (2.15) of the material filling the half-space.

As in the case of thin layers, we can now introduce rational approximations and find approximate boundary conditions of desired accuracy.

3.1.2 Approximations

Let us begin the discussion of the approximate boundary condition from the classical simple impedance boundary condition often used to model interfaces with good conductors.

Leontovich Impedance Boundary Condition

If the medium is electrically very dense as compared to that in the region $z > 0$, that is, $|k| \gg |k_0|$, where k and k_0 are the wavenumbers in the media at $z < 0$ and $z > 0$, respectively. If the sources are in the half-space $z > 0$, the tangential components of the wave vectors of all *propagating* plane-wave components satisfy the inequality $|\mathbf{k}_t| \leq |k_0|$. Thus, in the situation when the interface is in the far zone of the sources, we can assume that $|\mathbf{k}_t| \ll |k|$, meaning that the tangential derivatives in (3.6) can be neglected. This way we arrive at the standard Leontovich boundary condition

$$\mathbf{E}_{t+} = \eta\, \mathbf{n} \times \mathbf{H}_{t+} \qquad (3.7)$$

This simple condition is very widely used, and it is important to understand the limitations of the model. The limitations can be most conveniently formulated in terms of the *skin depth*. Indeed, the adopted approximation implies that waves inside the modeled medium travel nearly normally to the interface and the wave vector is very large. In other words, the fields in the medium vary very quickly; most commonly, if the material is a conductor, they actually quickly decay. We denote by δ

$$\delta = \frac{1}{\text{Im}\{k\}} \qquad (3.8)$$

the skin depth in metal. For good conductors

$$\delta = \sqrt{\frac{2}{\omega\sigma\mu}} \qquad (3.9)$$

where σ is the metal conductivity. If the material is electrically dense but the conductivity is low, the same meaning in the present contents has

$$\delta = \frac{1}{\text{Re}\{k\}} \qquad (3.10)$$

as a measure of the distance at which the field changes substantially. The approximation leading to (3.7) assumes that the variations *along* the interface are much slower compared to the variations in the normal direction. Thus, the following restrictions should be observed (here we formulate them for the case of a metal making an interface with free space). First, the condition $|k| \gg k_0 = \omega \sqrt{\epsilon_0 \mu_0}$ must be satisfied. In terms of the skin depth,

$$\delta \ll \frac{\lambda_0}{2\pi} \tag{3.11}$$

where $\lambda_0 = 2\pi/k_0$ is the free-space wavelength. Suppose that the interface is excited by a spherical wave with the curvature radius R. Then there is an additional condition

$$\delta \ll R \tag{3.12}$$

In particular, this means that the Leontovich condition cannot be used if sources are positioned very near to the modeled interface. Evanescent modes near sources have very large tangential propagation constants, and the condition $|\mathbf{k}_t| \ll |\mathbf{k}|$ is not satisfied.[1]

If our sample is a metal slab of thickness D, then we have the restriction

$$\delta \ll D \tag{3.13}$$

Finally, if the metal body is curved with the curvature radius a, we demand that

$$\delta \ll a \tag{3.14}$$

To estimate the error we can compare the solutions for the plane-wave reflection problem from a planar interface. Solving this problem exactly and using the Leontovich condition, we find that the results look similar: In the approximate solution we get $n = \sqrt{(\epsilon\mu)/(\epsilon_0\mu_0)}$ in place of $\sqrt{n^2 - \sin^2\theta}$ (θ is the incidence angle). Thus, the error in determining the reflection coefficient is of the order of $1/n^2$.

Higher-Order Conditions

To improve the model accuracy, we must introduce a rational approximation of the square root in (3.6). If $|k| \gg |\mathbf{k}_t|$ is valid, we may use the Taylor expansion

$$\frac{1}{\sqrt{k^2 + \nabla_t^2}} \approx \frac{1}{k}\left(1 - \frac{\nabla_t^2}{2k^2}\right) \tag{3.15}$$

and neglect the fourth-order (and higher) derivatives. This leads to the second-order boundary condition

$$\mathbf{E}_{t+} = \eta\left(\overline{\overline{I}}_t - \frac{\nabla_t^2}{2k^2} + \frac{\nabla_t\nabla_t}{k^2}\right) \cdot \mathbf{n} \times \mathbf{H}_{t+} \tag{3.16}$$

---

[1] More about evanescent modes and their decay can be found in Chapter 4.

Alternatively, we can multiply (3.6) by the denominator and again use the Taylor expansion. This results in a boundary condition containing second-order derivatives of both electric and magnetic fields:

$$\left(1 + \frac{\nabla_t^2}{2k^2}\right) \mathbf{E}_{t+} = \eta \left(\overline{\overline{I}}_t + \frac{\nabla_t \nabla_t}{k^2}\right) \cdot \mathbf{n} \times \mathbf{H}_{t+} \qquad (3.17)$$

A similar idea was used in [5] in studies of thin layers.

Other approximations can be useful in other circumstances. If the material contrast is not very high, the tangential derivatives of the fields are not necessarily small compared to the total spatial derivative inside the medium half-space.[2] Obviously, the Taylor expansion around zero $\mathbf{k}_t$ is not a reasonable approximation under these conditions, and it is better to use a more uniform rational approximation. Suppose we model an interface between free space (free-space wavenumber k_0) and a medium with the wavenumber k. For plane waves we can introduce a rational approximation

$$\sqrt{1 - \frac{k_t^2}{k^2}} \approx a - b\frac{k_t^2}{k^2} \qquad (3.18)$$

where we are free to choose constant parameters a and b. For example, we can demand that the approximation give exact results for waves at the normal incidence ($k_t = 0$) and for the 45° incidence angle[3] ($k_t = k_0/\sqrt{2}$). This determines $a = 1$ and

$$b = \frac{2k^2 - \sqrt{2}\,k\sqrt{2k^2 - k_0^2}}{k_0^2} \qquad (3.19)$$

The corresponding boundary condition reads

$$\left(1 + b\frac{\nabla_t^2}{k^2}\right) \mathbf{E}_{t+} = \eta \left(\overline{\overline{I}}_t + \frac{\nabla_t \nabla_t}{k^2}\right) \cdot \mathbf{n} \times \mathbf{H}_{t+} \qquad (3.20)$$

Note that this last result depends on the material *outside* the half-space that we model (the medium at $z > 0$ that we have assumed to be free space).

Similarly, we can demand that the approximation[4]

$$\frac{1}{\sqrt{k^2 - k_t^2}} \approx \frac{1}{k}\left(1 + b\frac{k_t^2}{k^2}\right) \qquad (3.21)$$

be valid exactly at $k_t = k_0/\sqrt{2}$, which corresponds to

$$b = \frac{k^2}{k_0^2}\left(\frac{k\sqrt{2}}{\sqrt{k^2 - k_0^2/2}} - 1\right) \qquad (3.22)$$

---

[2]In the formal language, the inequality $|\mathbf{k}_t| \ll |k|$ is not true.
[3]Or at some other angle, which might be more important in a specific situation.
[4]Compare with (3.15).

The boundary condition is then

$$\mathbf{E}_{t+} = \eta \left(\overline{\overline{I}}_t - b\frac{\nabla_t^2}{k^2} + \frac{\nabla_t \nabla_t}{k^2} \right) \cdot \mathbf{n} \times \mathbf{H}_{t+} \tag{3.23}$$

Also, the Padé approximation can be used for a more uniform model. The Padé approximation of order (2,2) reads

$$\sqrt{1 - \frac{k_t^2}{k^2}} \approx \frac{1 - \frac{3}{4}\frac{k_t^2}{k^2}}{1 - \frac{1}{4}\frac{k_t^2}{k^2}} \tag{3.24}$$

which leads to the boundary conditions in the form

$$\left(1 + \frac{3}{4}\frac{\nabla_t^2}{k^2} \right) \mathbf{E}_{t+} = \eta \left(\overline{\overline{I}}_t + \frac{1}{4}\frac{\nabla_t^2}{k^2} + \frac{\nabla_t \nabla_t}{k^2} \right) \cdot \mathbf{n} \times \mathbf{H}_{t+} \tag{3.25}$$

Chebyshev and other polynomial approximations are also possible; see more on this in Section 3.2.

Higher-order boundary conditions can be derived introducing higher-order approximations (e.g., the fourth order)

$$\sqrt{1 - \frac{k_t^2}{k^2}} \approx a - b\frac{k_t^2}{k^2} + c\frac{k_t^4}{k^4} \tag{3.26}$$

Using this formula we can get boundary conditions that are exact for *three* arbitrarily chosen incidence angles.

3.1.3 Numerical Examples: Accuracy of Approximate Solutions

To illustrate the accuracy of various approximate boundary conditions we consider a specific example of plane waves. For plane incident waves we use the Fourier domain boundary conditions to calculate the reflection coefficients. For TM polarized waves (see Figure 2.7), the tangential component of the electric field at the interface is

$$E_t = E_0(1 + R) \cos \theta \tag{3.27}$$

where E_0 is the amplitude of the incident electric field, and θ is the incidence angle measured from the normal direction. Vector $\mathbf{E}_t$ is directed along $\mathbf{k}_t$. The tangential magnetic field amplitude is

$$H_t = \frac{E_0}{\eta_0}(1 - R) \tag{3.28}$$

and vector $\mathbf{n} \times \mathbf{H}_t$ is also directed along $\mathbf{k}_t$. After substitution of these field amplitudes and $k_t = k_0 \sin \theta$ into the boundary conditions (3.5) (the exact solution), (3.16), and (3.17), the reflection coefficient is determined.

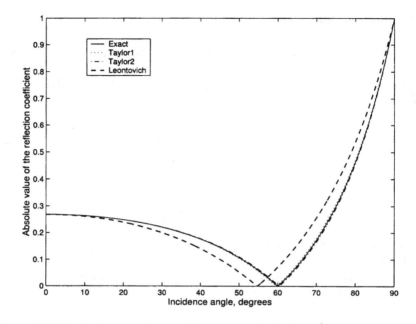

Figure 3.2 Reflection coefficient (absolute values) for a TM-polarized plane wave incident on a planar interface with the Earth's surface. Different approximate models compared with the exact solution. The curve marked "Taylor1" corresponds to (3.16), and label "Taylor2" corresponds to (3.17).

For TE polarized waves (Figure 2.8), we have

$$E_t = E_0(1 + R), \quad \text{and} \quad H_t = \frac{E_0}{\eta_0}(1 - R)\cos\theta \qquad (3.29)$$

and now both $\mathbf{E}_t$ and $\mathbf{n} \times \mathbf{H}_t$ are along $\mathbf{n} \times \mathbf{k}_t$. The reflection coefficient can now be solved in the same way as for TM polarized fields.

Let us choose as an example a plane interface with the Earth's surface. For low humidity of the ground, the relative permittivity at microwave frequencies is approximately

$$\epsilon_r = 3 - j10^{-5}/(\omega\epsilon_0) \qquad (3.30)$$

(ω is the frequency and the conductivity is approximately 10^{-5} S/m). Figures 3.2 and 3.3 show calculated results for the absolute values of the reflection coefficients for the incident field at the frequency of 1 GHz. We observe that already second-order impedance boundary conditions based on the Taylor approximation of the square-root function work quite well for all incidence angles. Improvement of the model accuracy as compared with the simple Leontovich boundary condition is very noticeable.

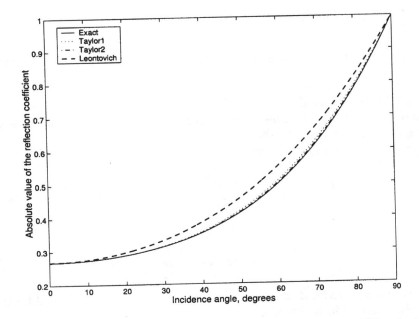

Figure 3.3 Reflection coefficient as in Figure 3.2 but for a TE-polarized incident plane wave.

3.2 APPLICATIONS TO NUMERICAL TECHNIQUES: ABSORBING BOUNDARY CONDITIONS

In the finite-difference numerical methods we need to truncate the computational domain with a "boundary" that simulates the absence of any boundary. This means that the truncation boundary should not reflect electromagnetic waves. Such boundaries can be approximately simulated by so-called *absorbing boundary conditions*.[5] These conditions can be obtained as special cases of the models of an interface in the case when the parameters of both media are the same.

Let us consider a planar "interface" between two vacuum half-spaces. Clearly, there is actually no boundary, but formally we can still apply the exact equivalent impedance conditions (3.5) and (3.6) to simulate the properties of one of the half-spaces. We only should substitute the free space

---

[5]Better results can be obtained using the perfectly matching layer technique, which is, however, more computationally demanding. The PML technique is considered in Section 6.1.2.

parameters η_0 and k_0 of the second half-space:

$$\mathbf{E}_{t+} = \overline{\overline{Z}} \cdot \mathbf{n} \times \mathbf{H}_{t+} \qquad \text{with} \qquad \overline{\overline{Z}} = \eta_0 \frac{\left(\overline{\overline{I}}_t - \frac{\mathbf{k}_t \mathbf{k}_t}{k_0^2}\right)}{\sqrt{1 - \frac{k_t^2}{k_0^2}}} \qquad (3.31)$$

If the fields will satisfy this boundary condition, there will be no reflection from the boundary on which the condition is set. In other words, after a formal transformation to the physical space, this corresponds to an exact *impedance absorbing boundary condition*:

$$\mathbf{E}_{t+} = \eta_0 \frac{\left(\overline{\overline{I}}_t + \frac{\nabla_t \nabla_t}{k_0^2}\right)}{\sqrt{1 + \frac{\nabla_t^2}{k_0^2}}} \cdot \mathbf{n} \times \mathbf{H}_{t+} \qquad (3.32)$$

or

$$\sqrt{1 + \frac{\nabla_t^2}{k_0^2}}\, \mathbf{E}_{t+} = \eta_0 \left(\overline{\overline{I}}_t + \frac{\nabla_t \nabla_t}{k_0^2}\right) \cdot \mathbf{n} \times \mathbf{H}_{t+} \qquad (3.33)$$

The diagonal dyadic $\overline{\overline{Z}}$ in (3.31) can be expressed in the orthogonal basis of its eigenvectors $\mathbf{k}_t$ and $\mathbf{n} \times \mathbf{k}_t$ as

$$\overline{\overline{Z}} = \eta_0 \left(\sqrt{1 - \frac{k_t^2}{k_0^2}} \frac{\mathbf{k}_t \mathbf{k}_t}{k_t^2} + \frac{1}{\sqrt{1 - \frac{k_t^2}{k_0^2}}} \frac{\mathbf{n} \times \mathbf{k}_t \mathbf{n} \times \mathbf{k}_t}{k_t^2}\right) \qquad (3.34)$$

simply by writing the two-dimensional unit dyadic in that basis:

$$\overline{\overline{I}}_t = \mathbf{x}_0 \mathbf{x}_0 + \mathbf{y}_0 \mathbf{y}_0 = \frac{\mathbf{k}_t \mathbf{k}_t}{k_t^2} + \frac{\mathbf{n} \times \mathbf{k}_t \mathbf{n} \times \mathbf{k}_t}{k_t^2} \qquad (3.35)$$

Equation (3.34) gives an exact expression for the input impedance of a half-space filled by vacuum as seen by plane electromagnetic waves. Of course, it is simply equal to the wave impedance dyadic of free space (2.16), so the interface is matched. In the coordinate form (the axis z is orthogonal to the virtual interface), (3.31) reads

$$k_0 \sqrt{k_0^2 - k_x^2 - k_y^2}\, E_x = \eta_0 \left[-(k_0^2 - k_x^2)H_y - k_x k_y H_x\right] \qquad (3.36)$$

$$k_0 \sqrt{k_0^2 - k_x^2 - k_y^2}\, E_y = \eta_0 \left[(k_0^2 - k_y^2)H_x + k_x k_y H_y\right] \qquad (3.37)$$

3.2.1 Impedance Operator and Engquist-Majda Operator

The impedance boundary condition contains tangential components of both electric and magnetic fields at an interface. If we deal with an imaginary interface between two identical media as in (3.31), the impedance boundary condition can be rewritten in a different form, so that the condition involves the tangential component of only one of the fields. Let us transform (3.31) so that it would contain only the electric field vector. Recall that for an isotropic medium the Maxwell equations can be written in a form resembling the transmission-line equations (2.11). These equations connect the field components orthogonal to the unit vector $\mathbf{n}$, the unit vector along axis z. In our present case we naturally choose the vector $\mathbf{n}$ to be orthogonal to the boundary. Equation (2.10) expresses the normal derivative of the tangential magnetic field in terms of the tangential electric field. This suggests differentiation of relation (3.31) with respect to z and a substitution from (2.10), which gives[6]

$$\frac{\partial}{\partial z}\mathbf{E}_{t+} = \overline{\overline{Z}} \cdot \frac{\partial}{\partial z}\mathbf{n} \times \mathbf{H}_{t+} = \overline{\overline{Z}} \cdot \left[j\omega\epsilon_0 \mathbf{E}_{t+} - \frac{1}{j\omega\mu_0}\mathbf{k}_t \times (\mathbf{k}_t \times \mathbf{E}_{t+}) \right] \quad (3.38)$$

Next, we just insert the surface impedance operator $\overline{\overline{Z}}$ from (3.31), and after simple algebra arrive at

$$\frac{\partial}{\partial z}\mathbf{E}_{t+} = j\sqrt{k_0^2 - k_t^2}\,\mathbf{E}_{t+} = jk_0\sqrt{1 - \frac{k_t^2}{k_0^2}}\,\mathbf{E}_{t+} \quad (3.39)$$

The advantage of this form as compared with (3.31) is mainly due to the fact that the terms containing the vector operator $\mathbf{k}_t\mathbf{k}_t \leftrightarrow -\nabla_t\nabla_t$ cancel out, so that only scalar operators acting on *every separate* component of $\mathbf{E}_t$ remain. The price we have to pay is an additional differential operator with respect to z.

   Transforming relation (3.39) to the space-time domain formally replacing

$$j\omega \rightarrow \frac{\partial}{\partial t} = D_t \quad \text{and} \quad -k_t^2 \rightarrow \frac{\partial^2}{\partial x^2} + \frac{\partial^2}{\partial y^2} = D_x^2 + D_y^2 \quad (3.40)$$

(notations commonly used in the literature on the finite-difference time-domain method, e.g., [14]), we come to

$$\left[D_z - \frac{D_t}{c}\sqrt{1 + \frac{D_x^2 + D_y^2}{D_t^2/c^2}} \right]\mathbf{E}_{t+} = 0 \quad (3.41)$$

Here $c = 1/\sqrt{\epsilon_0\mu_0}$ is the speed of light. This is the well-known Engquist-Majda pseudodifferential operator [14, 15] usually derived by a formal factorization of the wave equation. Connection to the wave equation is indeed

---

[6]Note that although boundary conditions are valid only on the interface surface, in this particular case we can differentiate it with respect to z, because there is no actual boundary, and the relation that we differentiate is valid also behind and above the surface. In the general case (e.g., when modeling high-conductivity bodies), the following derivation would be incorrect.

obvious from (3.39), because

$$\left(j\sqrt{k_0^2 - k_t^2} - \frac{\partial}{\partial z} \right) \left(j\sqrt{k_0^2 - k_t^2} + \frac{\partial}{\partial z} \right) = -k_0^2 + k_t^2 - \frac{\partial^2}{\partial z^2} \rightarrow -(\nabla^2 + k_0^2)$$

(3.42)

As we see now, the Engquist-Majda operator is equivalent to the exact impedance operator, and in both cases the difficulty is in finding a suitable approximation for the square root $\sqrt{1 - \frac{k_t^2}{k_0^2}} \rightarrow \sqrt{1 + \frac{\nabla_t^2}{k_0^2}}$. This problem is discussed in the next section.

3.2.2   Absorbing Boundary Conditions

Similarly to the case of material interfaces considered before, the exact operational absorbing boundary conditions cannot be practically used, and the square root operator should be approximated by a rational function.

General Considerations Regarding Possible Approximations

When simulating highly conducting bodies, we made use of the fact that the absolute value of the wavenumber in metals k was much larger than that in free space (k_0), which meant that the transverse to the interface component of the wavenumber k_t was much smaller in the absolute value than $|k|$. This was so because $|k_t| \leq k_0 \ll |k|$, actually for propagating plane waves $|k_t| = k_0 \sin\theta$, where θ is the incidence angle. In the present situation when the two media are the same, the ratio k_t^2/k_0^2 can take any value from zero to one. Moreover, for the evanescent part of the source spectrum this ratio can be arbitrarily large. Approximations for small values of this ratio, like the Taylor expansion, are not quite appropriate if we want to arrive at an approximate absorbing boundary condition effective for (nearly) all incidence angles.

Once again, let us introduce a rational approximation for the square root: $\sqrt{1 - k_t^2/k_0^2} \approx A$. For TM-polarized plane waves, where vector $\mathbf{n} \times \mathbf{H}_t$ is along $\mathbf{k}_t$, the reflection coefficient from our fictitious boundary reads

$$R_{\text{TM}} = \frac{A - \sqrt{1 - x^2}}{A + \sqrt{1 - x^2}}$$

(3.43)

where we have denoted for short $x^2 = k_t^2/k_0^2 = \sin^2\theta$ ($0 \leq x \leq 1$). For the TE polarization we get

$$R_{\text{TE}} = \frac{\frac{1}{A} - \frac{1}{\sqrt{1-x^2}}}{\frac{1}{A} + \frac{1}{\sqrt{1-x^2}}} = -R_{\text{TM}}$$

(3.44)

We see that the accuracy of the approximation (for both polarizations) depends only on how accurately we approximate the square root $\sqrt{1 - x^2} =$

$\sin^2 \theta$ by a rational function. The approximation should be accurate within the range $0 \leq x \leq 1$, where zero corresponds to the normal incidence and unity corresponds to the propagation just along the interface.[7] A very reasonable choice seems to be

$$\sqrt{1 - x^2} \approx \frac{1 - x^2}{1 - ax^2} \tag{3.45}$$

where a is a constant parameter, because this approximate expression gives the exact values at both ends of the interval of our interest. Let us note, however, that when the incidence angle tends to 90°, that is, $x \to 1$, the reflection coefficient tends to plus or minus one, although the approximated value of $\sqrt{1 - x^2}$ tends to zero, which is the exact value. This is obvious from (3.43) and (3.44): $1 - x^2$ tends to zero faster than $\sqrt{1 - x^2}$. From this fact we conclude that it might be better to use a more general approximation

$$\sqrt{1 - x^2} \approx \frac{1 - bx^2}{1 - ax^2} \tag{3.46}$$

because we will anyway have $|R| = 1$ at $\theta = \pi/2$, but a proper choice of one extra free parameter b might improve performance at other angles. Moreover, if we can make a compromise assuming nonzero (but, of course, small) reflection at the normal incidence, an approximation with three free parameters a, b, d

$$\sqrt{1 - x^2} \approx \frac{d - bx^2}{1 - ax^2} \tag{3.47}$$

is appropriate.

Various Absorbing Boundary Conditions

In the following derivations,[8] we consider the two-dimensional TM_x-case with the fields constant along x ($k_x = 0$). The exact absorbing boundary condition in the spectral domain (3.36) takes the form

$$\sqrt{k_0^2 - k_y^2}\, E_x = -\eta_0 k_0 H_y \tag{3.48}$$

Choosing a rational approximation for the square root in the general form (3.47), we find an approximate second-order absorbing boundary condition in the Fourier domain:

$$(dk_0^2 - bk_y^2)E_x = -\eta_0(k_0^2 - ak_y^2)H_y \tag{3.49}$$

---

[7]Because we assume that $x < 1$, these absorbing boundaries can distort reactive fields of closely located objects.

[8]The material in this section is based on work of Mikko Kärkkäinen [16].

Using the relation $k_0^2 = \omega^2/c^2$, where $c = 1/\sqrt{\epsilon_0 \mu_0}$ is the speed of light, with the Fourier-transform pairs $j\omega \leftrightarrow \partial/\partial t$ and $-jk_y \leftrightarrow \partial/\partial y$, we obtain the partial differential equation

$$\frac{d}{c^2}\frac{\partial^2 E_x}{\partial t^2} - b\frac{\partial^2 E_x}{\partial y^2} = -\frac{\eta_0}{c^2}\frac{\partial^2 H_y}{\partial t^2} + \eta_0 a \frac{\partial^2 H_y}{\partial y^2} \tag{3.50}$$

It is convenient to express the second-order time derivative of H_y in terms of the electric field component E_x:

$$\frac{\partial^2 H_y}{\partial t^2} = -\frac{c}{\eta_0}\frac{\partial^2 E_x}{\partial z \partial t} \tag{3.51}$$

The resulting equation is

$$\frac{d}{c^2}\frac{\partial^2 E_x}{\partial t^2} - b\frac{\partial^2 E_x}{\partial y^2} = \frac{1}{c}\frac{\partial^2 E_x}{\partial z \partial t} + \eta_0 a \frac{\partial^2 H_y}{\partial y^2} \tag{3.52}$$

We discretize this equation about an auxiliary lattice point, located one half-cell away from the interface. Note that we do not have to neglect any spatial or temporal differences. The resulting update equation for the electric field is

$$
\begin{aligned}
E_x|_{i,0}^{n+1} = & \; -E_x|_{i,1}^{n-1} + \frac{2d\Delta z}{c\Delta t + d\Delta z}\left(E_x|_{i,0}^n + E_x|_{i,1}^n\right) \\
& + \frac{c\Delta t - d\Delta z}{c\Delta t + d\Delta z}\left(E_x|_{i,1}^{n+1} + E_x|_{i,0}^{n-1}\right) \\
& + \frac{b(c\Delta t)^2 \Delta z}{\Delta y^2(c\Delta t + d\Delta z)}\left(\begin{array}{c} E_x|_{i+1,1}^n - 2E_x|_{i,1}^n + E_x|_{i-1,1}^n \\ +E_x|_{i+1,0}^n - 2E_x|_{i,0}^n + E_x|_{i-1,0}^n \end{array}\right) \\
& + \frac{\eta_0 a(c\Delta t)^2 \Delta z}{\Delta y^2(c\Delta t + d\Delta z)}\left(\begin{array}{c} H_y|_{i+1,1/2}^{n+1/2} - 2H_y|_{i,1/2}^{n+1/2} + H_y|_{i-1,1/2}^{n+1/2} \\ +H_y|_{i+1,1/2}^{n-1/2} - 2H_y|_{i,1/2}^{n-1/2} + H_y|_{i-1,1/2}^{n-1/2} \end{array}\right)
\end{aligned}
\tag{3.53}
$$

We notice that in special cases $d = 1, a = b = 0$, and $d = 1, b = 1/2, a = 0$, we have the first- and the second-order Mur absorbing boundary conditions [17], respectively. These conditions are based on a rather coarse approximation of the square root in equation (3.47). The first-order Mur condition corresponds to $\sqrt{1 - x^2} \approx 1$, and the second-order Mur condition is obtained for $\sqrt{1 - x^2} \approx 1 - x^2/2$ (Taylor series approximation). By choosing the coefficients of the rational approximation appropriately, it is possible to obtain much better absorption than in the second-order Mur conditions, while having essentially similar complexity of the update equation [18].

Usually, the third-order absorbing boundary conditions resulting from the rational approximation of the square root in the form (3.47) are formulated as third-order partial differential equations for only one field component (they can be derived from the Engquist-Majda operator (3.41) making approximations of the square root). The use of the impedance boundary condition with both electric and magnetic field components allows us to develop FDTD schemes of the second order with similar performance as the usual third-order absorbing boundary conditions.

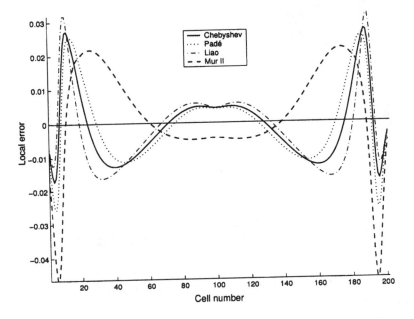

Figure 3.4 Local error on the absorbing boundary surface at time step $n = 150$.

To study the performance of the absorbing boundary conditions defined by (3.53), we have constructed a two-dimensional test lattice with the size of 20×200 cells. The source is a hard source at the center of the lattice with the time dependence

$$E_x|_{20,100}^n = \begin{cases} \frac{1}{32} [10 - 15 \cos(2\pi f n \Delta t) \\ +6 \cos(4\pi f n \Delta t) - \cos(6\pi f n \Delta t)], & n = 1, 2, \ldots 30 \\ 0, & n > 30 \end{cases}$$

(3.54)

where $f = 1$ GHz, $\Delta t = 0.9999 \Delta y/(\sqrt{2}c)$, and $\Delta y = \Delta z = 0.015$ m. This pulse has a very smooth decay to zero. The local error (difference between the incident field and the field at the absorbing boundary) calculated at time step $n = 150$ is shown in Figure 3.4. It is clear that the results of (3.53) corresponding to the Padé approximation ($d = 1$, $b = 0.75$, $a = 0.25$) and the Chebyshev ($d = 0.99973$, $b = 0.80864$, $a = 0.31657$) approximation of the square root in (3.47) are much better than the second-order Mur condition. Actually, the performance is seen to be about as good as that of the third-order Liao absorbing boundary condition [19].

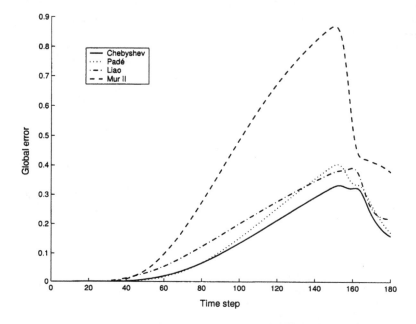

Figure 3.5 The global error (squared errors summed over the mesh width) as a function of time.

Let us next consider the global error, which probably gives a better measure for the performance of the absorbing conditions, since the squared errors are calculated and summed over the whole width of the boundary. The global error as a function of time steps is shown in Figure 3.5. It is evident from the figure that the second-order impedance conditions perform much better than the second-order Mur conditions. In the case of the Chebyshev approximation, the global error is seen to be even smaller than for the Liao third-order scheme [19]. The decay of the global error after about $n = 150$ just reflects the fact that the source has gone to zero some time ago; thus, the errors also become smaller.

3.3   HIGHER-ORDER MODELS OF THIN LAYERS

Next we introduce higher-order transition conditions for isotropic material layers. With the use of these results it is possible to avoid computing the fields inside the layer, properties of which are approximately modeled by certain relations between the tangential fields and their tangential derivatives defined on the two boundaries of the layer. The approximate relations can be found from suitable approximations of the exact boundary conditions in the Fourier domain (Section 2.3).

3.3.1 Thin Layers with $|\beta|d \ll 1$

Let us first study layers whose thickness is small compared with the wavelength in the direction orthogonal to the interfaces. Physically this means that the tangential fields change little from one side to the other side of the layer, and mathematically this means that $|\beta|d = |\sqrt{k^2 - k_t^2}|d \ll 1$, where d is the slab thickness. In practice this can refer to two different situations. First, we can have both $|k|d \ll 1$ and $|k_t|d \ll 1$, meaning that the slab is thin compared to the wavelength in the slab material and also to the wavelength of waves traveling along the layer. Second, the same assumption is valid if $|k_t| \approx |k|$, for example, for waves along a multilayer structure with small differences between parameters of layers, even if the layers are not electrically thin.

Let us use the exact transition conditions for an isotropic slab (2.111), and expand the coefficients in that relation in Taylor series with respect to the small parameter $|\beta|d$ retaining the second-order terms, which gives

$$\overline{\overline{a}}_{11} = \overline{\overline{a}}_{22} \approx \left[1 - \frac{(\beta d)^2}{2} \right] \overline{\overline{I}}_t \tag{3.55}$$

$$\overline{\overline{a}}_{12} \approx j\eta k d \left[1 - \frac{(\beta d)^2}{12} \right] \overline{\overline{A}} \tag{3.56}$$

$$\overline{\overline{a}}_{21} \approx \frac{j}{\eta} k d \left[1 - \frac{(\beta d)^2}{12} \right] \overline{\overline{C}} \tag{3.57}$$

This can be transformed to the physical space:

$$\mathbf{E}_{t+} = \left[1 - \frac{d^2}{2} (k^2 + \nabla_t^2) \right] \cdot \mathbf{E}_{t-}$$

$$+ j\eta k d \left[\left(1 - \frac{(kd)^2}{12} \right) \overline{\overline{I}}_t - \frac{d^2}{12} \nabla_t^2 + \frac{\nabla_t \nabla_t}{k^2} \right] \cdot \mathbf{n} \times \mathbf{H}_{t-} \tag{3.58}$$

$$\mathbf{n} \times \mathbf{H}_{t+} = \frac{j}{\eta} k d \left[\left(1 - \frac{(kd)^2}{12} \right) \overline{\overline{I}}_t - \frac{d^2}{12} \nabla_t^2 + \frac{\mathbf{n} \times \nabla_t \mathbf{n} \times \nabla_t}{k^2} \right] \cdot \mathbf{E}_{t-}$$

$$+ \left[1 - \frac{d^2}{2} (k^2 + \nabla_t^2) \right] \cdot \mathbf{n} \times \mathbf{H}_{t-} \tag{3.59}$$

Neglecting the terms proportional to d^2 and d^3 reduces these approximate conditions to the result of the locally quasistatic model given by (2.34) and (2.35).

Alternative second-order transition conditions can be derived from other vector circuit models. From the Thévenin equivalent voltage source representation (2.138) we get, assuming $|\beta| d \ll 1$,

$$\left[1 - \frac{d^2}{2}(k^2 + \nabla_t^2)\right] \cdot \mathbf{E}_{t+}$$

$$= \mathbf{E}_{t-} + j\eta kd\left[\left(1 - \frac{(kd)^2}{6}\right)\overline{\overline{I}}_t - \frac{d^2}{6}\nabla_t^2 + \frac{\nabla_t\nabla_t}{k^2}\right] \cdot \mathbf{n} \times \mathbf{H}_{t+} \qquad (3.60)$$

and from the Norton circuit model,

$$\left[1 - \frac{d^2}{2}(k^2 + \nabla_t^2)\right] \cdot \mathbf{n} \times \mathbf{H}_{t+}$$

$$= \mathbf{n} \times \mathbf{H}_{t-} + j\frac{kd}{\eta}\left[\left(1 - \frac{(kd)^2}{6}\right)\overline{\overline{I}}_t - \frac{d^2}{6}\nabla_t^2 + \frac{\mathbf{n} \times \nabla_t\mathbf{n} \times \nabla_t}{k^2}\right] \cdot \mathbf{E}_{t+} \quad (3.61)$$

To find approximate conditions for slabs backed by ideal electric and magnetic walls from (3.60) and (3.61), it is enough to let $\mathbf{E}_{t-} = 0$ for an ideally conducting surface or $\mathbf{H}_{t-} = 0$ for a magnetic wall.

3.3.2   Slowly Varying Fields Along the Interfaces

Next, let us assume that the propagation constant in the direction tangential to the slab is much smaller compared with the wavenumber k inside the slab material:

$$|\mathbf{k}_t| \ll |k| \qquad (3.62)$$

This can be valid if the refractive index of the layer material is large. If the layer is in free space, this means $|\epsilon\mu| \gg \epsilon_0\mu_0$. In more general cases, the comparison should be made with the properties of the surrounding media. In particular, condition (3.62) is not valid if there are other layers of materials with high refractive indices near the slab.

Another practical situation that refers to (3.62) is when the slab is excited by an external field varying slowly along the transversal plane. As a special case, assumption (3.62) is obviously valid for normal plane-wave excitation, when the transverse wavenumber $\mathbf{k}_t = 0$.

Other forms of the approximate boundary conditions are needed to model these situations. To derive appropriate boundary conditions, we express the exact equivalent parameters of the layer neglecting the terms of the order $(k_t/k)^2$ and higher. This leaves us with the following transition conditions:

$$\mathbf{E}_{t+} = \cos kd\,\mathbf{E}_{t-} + j\eta\sin kd\,\mathbf{n} \times \mathbf{H}_{t-} \qquad (3.63)$$

$$\mathbf{n} \times \mathbf{H}_{t+} = \cos kd\,\mathbf{n} \times \mathbf{H}_{t-} + \frac{j}{\eta}\sin kd\,\mathbf{E}_{t-} \qquad (3.64)$$

In the same way from the Thévenin and Norton models (2.138) and (2.139), we find

$$\mathbf{E}_{t+} = \frac{1}{\cos kd}\,\mathbf{E}_{t-} + j\eta \tan kd\,\mathbf{n} \times \mathbf{H}_{t+} \tag{3.65}$$

$$\mathbf{n} \times \mathbf{H}_{t+} = \frac{1}{\cos kd}\,\mathbf{n} \times \mathbf{H}_{t-} + \frac{j}{\eta}\tan kd\,\mathbf{E}_{t+} \tag{3.66}$$

These very simple conditions can be effectively used in the frequency domain. Finally, if the slab is thin in wavelengths:

$$|k|d \ll 1 \tag{3.67}$$

and condition (3.62) also holds, we can expand the trigonometric functions and still simplify the model:

$$\mathbf{E}_{t+} = \left[1 - \frac{(kd)^2}{2}\right]\mathbf{E}_{t-} + j\eta kd\,\mathbf{n} \times \mathbf{H}_{t-} \tag{3.68}$$

$$\mathbf{n} \times \mathbf{H}_{t+} = \left[1 - \frac{(kd)^2}{2}\right]\mathbf{n} \times \mathbf{H}_{t-} + \frac{j}{\eta}kd\,\mathbf{E}_{t-} \tag{3.69}$$

(second-order terms have been retained). Under the same assumptions we also get

$$\left[1 - \frac{(kd)^2}{2}\right]\mathbf{E}_{t+} = \mathbf{E}_{t-} + j\eta kd\,\mathbf{n} \times \mathbf{H}_{t+} \tag{3.70}$$

$$\left[1 - \frac{(kd)^2}{2}\right]\mathbf{n} \times \mathbf{H}_{t+} = \mathbf{n} \times \mathbf{H}_{t-} + \frac{j}{\eta}kd\,\mathbf{E}_{t+} \tag{3.71}$$

The last results are also suitable for time-domain modeling, since the transformation to the time domain can now be made, simply replacing

$$jk \rightarrow \sqrt{\epsilon_0\mu_0}\,\frac{\partial}{\partial t} \tag{3.72}$$

Note that this approximation for thin layers is based on the same physical assumption as the Leontovich boundary condition for a half-space, due to the assumption that *inside* the layer the waves travel approximately in the direction normal to the interfaces. An equation, similar to (3.68) with $\mathbf{E}_{t-} = 0$, appeared probably for the first time in 1959 in [20]. This approximation was extended for thin ferrite layers on ideally conducting surfaces in [21].

If neither of $|\beta|d$ and $|k_t|d$ is small, higher-order boundary conditions can still be derived making polynomial approximations of functions $\cos\beta d$ and $\sin(\beta d)/\beta$, which come into the exact expressions for the equivalent circuit parameters of isotropic layers (2.112) and (2.113). For example, we can demand the approximation give exact values for two or more incident angles [5]. Padé and Chebyshev approximations can be applied as well, similarly to models of interfaces with half-spaces (Sections 3.1.2 and 3.2.2).

3.4 HIGHER-ORDER IMPEDANCE BOUNDARY CONDITIONS FOR LAYERS ON METAL SURFACES

In Chapter 2, the following exact boundary condition (2.143) for an isotropic layer of thickness d on an ideally conducting surface was derived:

$$\mathbf{E}_{t+} = j\frac{\eta k}{\beta}\tan(\beta d)\,\overline{\overline{A}}\cdot(\mathbf{n}\times\mathbf{H}_{t+}) \tag{3.73}$$

This relation is valid for plane waves, and we need to transform it to the physical space to get impedance boundary conditions:

$$\mathbf{E}_{t+} = j\frac{\eta k}{\sqrt{k^2+\nabla_t^2}}\tan\left(\sqrt{k^2+\nabla_t^2}\,d\right)\left(\overline{\overline{I}}_t + \frac{\nabla_t\nabla_t}{k^2}\right)\cdot(\mathbf{n}\times\mathbf{H}_{t+}) \tag{3.74}$$

Comparing the last equation with (3.6), we note that they are very similar, so the same ideas for approximations can be applied. In this case, we introduce rational approximations for

$$f(\nabla_t^2) = \frac{\tan\left(\sqrt{k^2+\nabla_t^2}\,d\right)}{\sqrt{k^2+\nabla_t^2}} \tag{3.75}$$

3.4.1 Exact Surface Impedance Operator

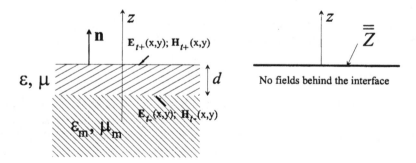

Figure 3.6 Layer of an isotropic material on the surface of a material half-space. Geometry and notations. Impedance model of the system is shown on the right.

Consider the situation when a material layer is used as a coating for a real metal surface, not an ideal conductor (Figure 3.6). We start the analysis from deriving an exact impedance operator, which can then be approximated as appropriate. This can be conveniently done making use of the vector circuit model for slabs (Section 2.3). A slab of an isotropic material is modeled by a dyadic transmission matrix as (2.111)

$$
\begin{pmatrix} \mathbf{E}_{t+} \\ \mathbf{n} \times \mathbf{H}_{t+} \end{pmatrix} = \begin{pmatrix} \overline{\overline{a}}_{11} & \overline{\overline{a}}_{12} \\ \overline{\overline{a}}_{21} & \overline{\overline{a}}_{22} \end{pmatrix} \cdot \begin{pmatrix} \mathbf{E}_{t-} \\ \mathbf{n} \times \mathbf{H}_{t-} \end{pmatrix}
\tag{3.76}
$$

with the dyadic coefficients given by (2.112) and (2.113):

$$
\overline{\overline{a}}_{11} = \overline{\overline{a}}_{22} = \cos(\beta d)\overline{\overline{I}}_t
$$

$$
\overline{\overline{a}}_{12} = j\frac{k\eta}{\beta} \sin(\beta d) \left(\overline{\overline{I}}_t - \frac{\mathbf{k}_t \mathbf{k}_t}{k^2} \right)
\tag{3.77}
$$

$$
\overline{\overline{a}}_{21} = j\frac{k}{\eta\beta} \sin(\beta d) \left(\overline{\overline{I}}_t - \frac{\mathbf{n} \times \mathbf{k}_t\, \mathbf{n} \times \mathbf{k}_t}{k^2} \right)
$$

As always, here $k = \omega\sqrt{\epsilon\mu}$, $\eta = \sqrt{\mu/\epsilon}$, and $\beta = \sqrt{k^2 - k_t^2}$ is the normal to the interface component of the wave vector in the slab.

Suppose now that the slab is located at an interface with a metal (or a half-space filled by another isotropic material). Let us mark the parameters of this second material as ϵ_m, μ_m, and use the same index for all other parameters of metal ($\eta_m = \sqrt{\mu_m/\epsilon_m}$, and so forth). Our first goal here is to derive the exact impedance boundary condition for the free surface of the material layer. As we know, the half-space filled by a lossy isotropic material can be modeled by an exact impedance operator (3.5), which connects the fields on the interface between metal and the dielectric layer:

$$
\mathbf{E}_{t-} = \frac{\eta_m k_m}{\beta_m} \left(\overline{\overline{I}}_t - \frac{\mathbf{k}_t \mathbf{k}_t}{k_m^2} \right) \cdot \mathbf{n} \times \mathbf{H}_{t-} = \overline{\overline{Z}}_m \cdot \mathbf{n} \times \mathbf{H}_{t-}
\tag{3.78}
$$

Next we substitute this relation into (3.76) and eliminate the fields on the intermediate interface $\mathbf{E}_{t-}$, $\mathbf{H}_{t-}$. In terms of the dyadic coefficients of (3.76), the result is

$$
\mathbf{E}_{t+} = \left(\overline{\overline{a}}_{11} \cdot \overline{\overline{Z}}_m + \overline{\overline{a}}_{12} \right) \cdot \left(\overline{\overline{a}}_{21} \cdot \overline{\overline{Z}}_m + \overline{\overline{a}}_{22} \right)^{-1} \cdot \mathbf{n} \times \mathbf{H}_{t+}
\tag{3.79}
$$

We see that all the dyadic quantities in (3.79), including $\overline{\overline{Z}}_m$, share the same basis of eigenvectors: the transverse wave vector $\mathbf{k}_t$ and the orthogonal vector $\mathbf{n} \times \mathbf{k}_t$. It is convenient to make actual calculations in this basis, making use of (2.114) and (2.115). The result reads

$$
\mathbf{E}_{t+} = \overline{\overline{Z}} \cdot \mathbf{n} \times \mathbf{H}_{t+}
\tag{3.80}
$$

with

$$
\overline{\overline{Z}} = \frac{\frac{\eta_m \beta_m}{k_m} \cos\beta d + j\frac{\eta\beta}{k} \sin\beta d}{\cos\beta d + j\frac{k}{\eta\beta}\frac{\eta_m\beta_m}{k_m} \sin\beta d} \frac{\mathbf{k}_t\mathbf{k}_t}{k_t^2} + \frac{\frac{\eta_m k_m}{\beta_m} \cos\beta d + j\frac{\eta k}{\beta} \sin\beta d}{\cos\beta d + j\frac{\beta}{\eta k}\frac{\eta_m k_m}{\beta_m} \sin\beta d} \frac{\mathbf{n} \times \mathbf{k}_t\, \mathbf{n} \times \mathbf{k}_t}{k_t^2}
\tag{3.81}
$$

As a check, we see that for $\beta d \to 0$ the result reduces to the exact impedance operator for a free metal surface (3.5). Written in this basis,

$$\overline{\overline{Z}} = \frac{\eta_m \beta_m}{k_m} \frac{\mathbf{k}_t \mathbf{k}_t}{k_t^2} + \frac{\eta_m k_m}{\beta_m} \frac{\mathbf{n} \times \mathbf{k}_t \, \mathbf{n} \times \mathbf{k}_t}{k_t^2} \tag{3.82}$$

In the limit $\eta_m \to 0$, that is, for an ideally conducting backing, (3.81) reduces to (2.143). In the basis of the eigenvectors, this reads

$$\overline{\overline{Z}} = j\frac{\eta\beta}{k} \tan \beta d \, \frac{\mathbf{k}_t \mathbf{k}_t}{k_t^2} + j\frac{\eta k}{\beta} \tan \beta d \, \frac{\mathbf{n} \times \mathbf{k}_t \, \mathbf{n} \times \mathbf{k}_t}{k_t^2} \tag{3.83}$$

Next, in the view of transforming the result from the Fourier space to the physical domain and deriving higher-order impedance boundary conditions, we transform the result into a more convenient form such as (3.5). To do so, we simply rewrite our diagonal dyadic in (3.81) as

$$a\frac{\mathbf{k}_t \mathbf{k}_t}{k_t^2} + b\frac{\mathbf{n} \times \mathbf{k}_t \, \mathbf{n} \times \mathbf{k}_t}{k_t^2} \quad \to \quad b\overline{\overline{I}}_t + (a-b)\frac{\mathbf{k}_t \mathbf{k}_t}{k_t^2} \tag{3.84}$$

Calculating $a - b$, we see that it is proportional to k_t^2, so that k_t^2 cancels out in the last term. After some algebra we write (3.81) in an equivalent form

$$\overline{\overline{Z}} = \frac{\frac{\eta_m k_m}{\beta_m} \cos \beta d + j\frac{\eta k}{\beta} \sin \beta d}{\cos \beta d + j\frac{\beta}{\eta k} \frac{\eta_m k_m}{\beta_m} \sin \beta d} \overline{\overline{I}}_t$$

$$- \frac{\frac{\eta_m}{k_m \beta_m} \cos^2 \beta d + j\frac{\eta}{2k\beta}\left(1 + \frac{\eta_m^2}{\eta^2}\right) \sin 2\beta d - \left(\frac{1}{k^2} + \frac{1}{\beta^2}\right)\frac{\eta_m \beta_m}{k_m} \sin^2 \beta d}{\cos^2 \beta d + j\frac{\eta_m \beta_m}{2\eta k_m}\left(\frac{\beta}{k} + \frac{k}{\beta}\right) \sin 2\beta d - \frac{\eta_m^2}{\eta^2} \sin^2 \beta d} \mathbf{k}_t \mathbf{k}_t$$

$$\tag{3.85}$$

As a check, we see that for $\beta d \to 0$ (no covering layer), this simplifies to

$$\overline{\overline{Z}} = \frac{\eta_m k_m}{\beta_m}\left(\overline{\overline{I}}_t - \frac{\mathbf{k}_t \mathbf{k}_t}{k_m^2}\right) \tag{3.86}$$

which is the same as (3.5).

3.4.2 Thin Layers Covering Highly Conducting Bodies

If the metal surface can be modeled by the Leontovich boundary condition (3.7) (meaning that $\beta_m = \sqrt{k_m^2 - k_t^2}$ can be approximately replaced by k_m), the exact impedance operator (3.81) simplifies as

$$\overline{\overline{Z}} = \frac{\eta_m + j\frac{\eta\beta}{k} \tan \beta d}{1 + j\frac{k}{\eta\beta}\eta_m \tan \beta d} \frac{\mathbf{k}_t \mathbf{k}_t}{k_t^2} + \frac{\eta_m + j\frac{\eta k}{\beta} \tan \beta d}{1 + j\frac{\beta}{\eta k}\eta_m \tan \beta d} \frac{\mathbf{n} \times \mathbf{k}_t \, \mathbf{n} \times \mathbf{k}_t}{k_t^2} \tag{3.87}$$

Transforming (3.87) using (3.84), we get

$$\overline{\overline{Z}} = \frac{\eta_m + j\frac{\eta k}{\beta}\tan\beta d}{1 + j\frac{\beta}{\eta k}\eta_m\tan\beta d}\overline{\overline{I}}_t - \frac{j\frac{\eta}{k\beta}\left(1 + \frac{\eta_m^2}{\eta^2}\right)\tan\beta d - \eta_m\left(\frac{1}{\beta^2} + \frac{1}{k^2}\right)\tan^2\beta d}{1 + j\frac{\eta_m}{\eta}\left(\frac{\beta}{k} + \frac{k}{\beta}\right)\tan^2\beta d}\mathbf{k}_t\mathbf{k}_t$$

$$(3.88)$$

At this stage we are ready to make approximations and introduce higher-order impedance boundary conditions.

Approximations

Suppose at first that the material of the layer is electrically dense compared to the material filling the space at $z > 0$, meaning that $|k| \gg |k_t|$ or, equivalently, $\beta \approx k$. Making appropriate approximations we arrive at

$$\overline{\overline{Z}} = \frac{\eta_m + j\eta\tan kd}{1 + j\frac{\eta_m}{\eta}\tan kd}\left(\frac{\mathbf{k}_t\mathbf{k}_t}{k_t^2} + \frac{\mathbf{n}\times\mathbf{k}_t\,\mathbf{n}\times\mathbf{k}_t}{k_t^2}\right) = \frac{\eta_m + j\eta\tan kd}{1 + j\frac{\eta_m}{\eta}\tan kd}\overline{\overline{I}}_t \quad (3.89)$$

This is obviously just the formula for the impedance transformation along a transmission line of length d with the wave impedance η loaded by the impedance η_m. In this case the result is the same for both polarizations, so the input impedance dyadic reduces to a scalar. This is natural because in this approximation the waves in the slab and in the metal half-space are assumed to be traveling in the direction normal to the interfaces (since it has been assumed that $k_t \approx 0$). Formula (3.89) allows obvious simplifications if, for example, $|k|d \ll 1$.

Let us next assume that the restriction $k_t \ll |k|$ is not necessarily valid, but a particular system is such that there are two small parameters of the same order of smallness. First, the ratio of the wave impedance in metal η_m and that in the space above the layer η is small: $|\eta_m/\eta| \ll 1$. Second, the layer is thin in the sense that $|\beta|d \ll 1$. In this case (3.88) can be approximated as

$$\overline{\overline{Z}} = (\eta_m + j\eta kd)\overline{\overline{I}}_t - j\frac{\eta}{k}d\,\mathbf{k}_t\mathbf{k}_t \quad (3.90)$$

where only the first-order terms have been kept. This result can be interpreted as a sum of the "load impedance" η_m that is the surface impedance of the metal half-space, and an additional impedance due to the fields in the thin layer. If the layer is thin, the last part is proportional to βd. Transforming to the physical space we arrive at the approximate boundary condition

$$\mathbf{E}_{t+} = (\eta_m + j\eta kd)\,\mathbf{n}\times\mathbf{H}_{t+} + j\frac{\eta d}{k}\nabla_t\nabla_t\cdot(\mathbf{n}\times\mathbf{H}_{t+}) \quad (3.91)$$

PROBLEMS

3.1 For a planar interface with an isotropic medium, derive an "admittance" boundary condition in the form $\mathbf{n} \times \mathbf{H}_t = \overline{\overline{Y}} \cdot \mathbf{E}_t$ from (2.10). Check the result by calculating the two-dimensional inverse of dyadic $\overline{\overline{Z}}$ in (3.5).

3.2 Prove that if the Leontovich boundary condition is used to estimate plane-wave reflection coefficients from an interface between free space and a material, the error is of the order of $1/n^2$, where n is the refraction index of the material.

3.3 Study how the approximation coefficient in (3.20) depends on the ratio between the refractive indices of the two media $n = \sqrt{(\epsilon\mu)/(\epsilon_0\mu_0)}$. Compare (3.20) with (3.17). Explain the results based on physical reasoning.

3.4 Derive higher-order impedance boundary conditions for a magnetodielectric coating of an ideally conducting surface. Consider high- and low-contrast coatings.

3.5 Derive second-order approximate boundary conditions for a thin dielectric slab on a metal plane using the averaging method for the fields in the slab and the Leontovich impedance boundary condition to model the metal surface.

3.6 Study surface waves on an interface between two isotropic media. Derive the dispersion equation for TE and TM waves. Hint: Apply the exact impedance boundary conditions for two semi-infinite spaces.

References

[1] Shchukin, A.N., *Propagation of Radio Waves, Part I: Basic Theory of Wave Propagation over a Plane and a Sphere*, Moscow-Leningrad: Navy Publishers, USSR, 1940 (in Russian).

[2] Rytov, S.M., "Calcul du Skin-Effet par la Méthode des Perturbations," *J. Physics USSR*, Vol. 2, 1940, pp. 233-242 [Russian version in *Zhurnal Eksperimentalnoi i Teoreticheskoi Fiziki*, Vol. 10, 1940, pp. 180-189].

[3] Leontovich, M.A., *Investigations on Radio Wave Propagation, Part II*, Moscow: Academy of Sciences, 1948 (in Russian).

[4] Senior, T.B.A., and J.L. Volakis, *Approximate Boundary Conditions in Electromagnetics*, London: IEE Electromagnetic Waves Series, 1995.

[5] Hoppe, D.J., and Y. Rahmat-Samii, *Impedance Boundary Conditions in Electromagnetics*, Washington, D.C.: Taylor and Francis, 1995.

[6] Oksanen, M.I., S.A. Tretyakov, and I.V. Lindell, "Vector Circuit Theory for Isotropic and Chiral Slabs," *J. Electromagnetic Waves and Applications*, Vol. 4, No. 7, 1990, pp. 613-643.

[7] Lyalinov, M.A., and A.H. Serbest, "Transition Boundary Conditions for Simulation of Thin Chiral Slab," *Electronics Lett.*, Vol. 34, No. 12, 1998, pp. 1211-1213.

[8] Tretyakov, S.A., "Thin Pseudochiral Layers: Approximate Boundary Conditions and Potential Applications," *Microwave and Optical Technology Lett.*, Vol. 6, No. 2, 1993, pp. 112-115.

[9] Ricoy, M.A., and J.L. Volakis, "Derivation of Generalized Transition/Boundary Conditions for Planar Multi-Layer Structures," *Radio Science*, Vol. 25, 1990, pp. 391-405.

[10] Tretyakov, S.A., "Generalized Impedance Boundary Conditions for Isotropic Multilayers," *Microwave and Optical Technology Lett.*, Vol. 17, No. 4, 1998, pp. 262-265.

[11] Puska, P.P., S.A. Tretyakov, and A.H. Sihvola, "Approximate Impedance Boundary Conditions for Isotropic Multilayered Media," *IEE Proc. Microwave, Antennas and Propagation*, Vol. 146, No. 2, 1999, pp. 163-166.

[12] Cicchetti, R., "A Class of Exact and Higher-Order Surface Boundary Conditions for Layered Structures," *IEEE Trans. Antennas and Propagation*, Vol. 44, No. 2, 1996, pp. 249-259.

[13] Wait, J.R., *Geo-Electromagnetism*, New York: Academic Press, 1982.

[14] Taflove, A., and S.C. Hagness, *Computational Electrodynamics: The Finite-Difference Time-Domain Method*, 2nd ed., Norwood, MA: Artech House, 2000.

[15] Engquist, B., and A. Majda, "Absorbing Boundary Conditions for the Numerical Simulation of Waves," *Mathematics of Computation*, Vol. 31, 1977, pp. 629-651.

[16] Kärkkäinen, M.K., and S.A. Tretyakov, "A Class of Analytical Absorbing Boundary Conditions Originating From the Exact Surface Impedance Boundary Condition," *IEEE Trans. Microwave Theory and Techniques*, Vol. 51, No. 2, 2003, pp. 560-563.

[17] Mur, G., "Absorbing Boundary Conditions for the Finite-Difference Approximation of the Time-Domain Electromagnetic Field Equations," *IEEE Trans. Electromagnetic Compatibility*, Vol. 23, 1981, pp. 377-382.

[18] Moore, T.G., et al., "Theory and Application of Radiation Boundary Operators," *IEEE Trans. Antennas and Propagation,* Vol. 36, 1988, pp. 1797-1812.

[19] Liao, Z.P., et al., "A Transmitting Boundary for Transient Wave Analyses," *Acta Sinica* (Series A), Vol. XXVII, 1984, pp. 1063-1076.

[20] Armand, N.A., "Propagation of Surface Electromagnetic Waves Along a Multiconductor Line," *J. Technical Physics USSR,* Vol. 29, 1959, pp. 107-119 (in Russian).

[21] Kurushin, E.P., and E.I. Nefedov, *Electrodynamics of Anisotropic Waveguiding Structures,* Moscow: Nauka Press, 1983 (in Russian).

Chapter 4

Periodical Structures, Arrays, and Meshes

Various space-periodic structures find numerous applications in filters, antennas, and in the design of artificial materials. Periodical arrangements of planar conductive elements of various shapes are used as frequency selective surfaces. The most common numerical approach to the calculation of their characteristics is the periodical method of moments. Numerical models are not considered in this book; see, for example, [1, 2]. In this chapter, we will show how periodical structures can be modeled analytically starting from systems periodical only along one direction in space. We will begin with a general discussion of the properties of periodical waveguides, and then consider a simple and very illustrative example of a planar array of parallel conducting wires. Approximate boundary conditions will be derived for dense arrays and meshes of wires and strips. Next, models for periodical arrays of small dipole particles will be presented. In this chapter, we mainly consider models suitable for understanding and estimating reflection and transmission properties of periodical structures like wire or strip arrays and grids of small conducting particles. Wave phenomena and effective material properties of three-dimensional arrays and artificial media will be discussed in the next chapter. The fundamental concepts and models for the local field in regular arrays will be used in the following chapters, devoted to three-dimensional composites and applications.

4.1 FLOQUET MODES

First of all, we need to introduce the main concepts used in the description of various periodical structures. So-called Floquet harmonics will be explained next.

4.1.1   Floquet Modes in Periodical Waveguides

Let us consider a generic waveguide defined as a system whose electromagnetic properties do not change along at least one straight line in space. Now generalize this definition assuming that the waveguide parameters are not necessarily uniform along one straight line in space (defined by the Cartesian axis z) but can *periodically* depend on the z-coordinate. Of course, the regular waveguide is a special case of this more general class. Let us denote the period by d, then a periodical waveguide can be defined by its complex permittivity and permeability functions that satisfy the periodicity condition

$$\epsilon(x,y,z+d) = \epsilon(x,y,z), \qquad \mu(x,y,z+d) = \mu(x,y,z) \qquad (4.1)$$

for all z. Waveguides with metal walls (e.g., corrugated waveguides and slow-wave structures) are also covered by this definition as a limiting case with $\epsilon \to -j\infty$ in conducting regions.

Waves in periodical waveguides are governed by the Maxwell equations with material relations, which in this situation form a system of linear differential equations with *periodical coefficients*. We can make use of the Floquet theorem in the theory of differential equations to establish that the solutions in periodical waveguides satisfy relations

$$\mathbf{E}(x,y,z+nd) = e^{-jnqd}\,\mathbf{E}(x,y,z), \qquad \mathbf{H}(x,y,z+nd) = e^{-jnqd}\,\mathbf{H}(x,y,z)$$
$$(4.2)$$

Here $n = 0, \pm 1, \pm 2, \ldots$, and q is a complex constant. Relations (4.2) mean that the fields in the adjacent periods differ only by a constant complex multiplier. This observation allows us to write the fields at an arbitrary point as

$$\mathbf{E}(x,y,z) = \mathbf{E}_0(x,y,z)e^{-jqz}, \qquad \mathbf{H}(x,y,z) = \mathbf{H}_0(x,y,z)e^{-jqz} \qquad (4.3)$$

where the amplitude functions $\mathbf{E}_0$ and $\mathbf{H}_0$ in (4.3) are *periodical* functions of z with the period d. Thus, we can expand them in the Fourier series and write

$$\mathbf{E}(x,y,z) = \sum_{m=-\infty}^{\infty} \mathbf{E}_m(x,y)\, e^{-j\left(q+\frac{2\pi m}{d}\right)z} \qquad (4.4)$$

The members of this series are called *Floquet modes* or *spatial harmonics*.

In regular waveguides, functions $\mathbf{E}_0$ and $\mathbf{H}_0$ do not depend on z, and q is, obviously, the propagation factor. Periodical variations of the waveguide lead to a "modulation" of the amplitude functions: They become periodical functions with the same period d as that of the waveguide. Considering Floquet harmonics as waves along the waveguide axis z we note that the group velocity

$$v_{\mathbf{g}} = \frac{\partial \omega}{\partial \left(q + \frac{2\pi m}{d}\right)} \qquad (4.5)$$

is the same for all Floquet modes, because $2\pi m/d$ does not depend on the frequency. On the other hand, the phase velocities are different for different Floquet components:

$$v_{\text{ph}} = \frac{\omega}{q + \frac{2\pi m}{d}} \tag{4.6}$$

Every individual Floquet mode looks like an eigenwave in a regular waveguide

$$\mathbf{E}_m(x, y, z) = \mathbf{E}_m(x, y)\, e^{-jq_m z} \tag{4.7}$$

but taken separately a single Floquet mode has little physical meaning, because only the full sum of the Fourier series (4.4) satisfies both the Maxwell equations and the boundary conditions.

4.1.2 Example: Array of Parallel Current Lines

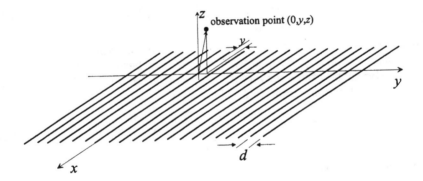

Figure 4.1 Planar regular array of thin conducting wires supporting currents I_n.

Consider a simple example of a periodical system: an array of infinite parallel thin ideally conducting wires that can be replaced by an array of line currents. Understanding of this system is important for the theory of antenna arrays and wire grids. All the wires belong to the plane (x, y), they are positioned at distance d apart, and each wire is parallel to the axis x (Figure 4.1). The wire radius is $r_0 \ll d$. Let us consider the electric field created by wire currents if an eigenwave propagates along the grid. We denote the currents in the wires as I_n (n is the number of the wire, and $n = 0$ corresponds to the wire at $y = 0$). They satisfy the Floquet theorem (4.2):

$$I_n = I_0\, e^{-jnqd} \tag{4.8}$$

provided that the excitation field (if there is any) is also a periodical function of y with the period d. For simplicity, we assume that there is no dependence of the currents and fields on the x-coordinate (along the wires).

Each wire current (say, the current in the wire number n) creates an electric field whose component in the direction along the wires (x-direction) is (see, e.g., [3, p. 492])

$$E_{xn} = -\frac{\eta k}{4} I_n H_0^{(2)} (k\, r_n) \qquad (4.9)$$

Here $r_n = \sqrt{z^2 + (nd - y)^2}$ is the distance from the wire axis to the observation point, $k = \omega\sqrt{\epsilon_0 \mu_0}$ is the wave number, $\eta = \sqrt{\mu_0/\epsilon_0}$ is the wave impedance, and $H_0^{(2)}$ is the Hankel function. In this model we assume that the wire radius is small compared to the wavelength ($k r_0 \ll 1$), and it is much smaller than the distance between the wires ($r_0 \ll d$). Because of the last assumption, we have replaced the wire field by the field of an infinitely thin current line along the wire axis. The field generated by the whole array at point $(0, y, z)$ is the sum of the fields of all wires:

$$E_x = -\frac{\eta k}{4} I_0 \sum_{n=-\infty}^{\infty} e^{-jnqd} H_0^{(2)} \left(k\sqrt{z^2 + (nd - y)^2} \right) \qquad (4.10)$$

Due to the grid periodicity, it is enough to study the field dependence on y only within one grid period, so we have assumed here that $0 < y < d$. This series of Hankel functions converges rather slowly. To improve its convergence we apply the Poisson summation rule:

$$\sum_{n=-\infty}^{\infty} f(nd) = \frac{1}{d} \sum_{m=-\infty}^{\infty} \int_{-\infty}^{\infty} f(\tau) e^{-j\frac{2\pi m}{d}\tau} \, d\tau \qquad (4.11)$$

In our case, substituting the Fourier transform of the Hankel function

$$\int_{-\infty}^{\infty} e^{-jt\tau} H_0^{(2)} \left(k\sqrt{z^2 + t^2} \right) dt = 2\frac{e^{-j\sqrt{k^2 - \tau^2}|z|}}{\sqrt{k^2 - \tau^2}}, \qquad \mathrm{Im}\{\sqrt{k^2 - \tau^2}\} < 0 \qquad (4.12)$$

we get

$$\sum_{n=-\infty}^{\infty} e^{-jnqd} H_0^{(2)} \left(k\sqrt{z^2 + (nd - y)^2} \right) = j\frac{2}{d} \sum_{m=-\infty}^{\infty} e^{-j\frac{2\pi my}{d}} \frac{e^{-\alpha_{mz}|z|}}{\alpha_{mz}} \qquad (4.13)$$

Here

$$\alpha_{mz} = \sqrt{\left(\frac{2\pi m}{d} + q\right)^2 - k^2} \qquad (4.14)$$

and the branch of the square root function is defined by $\mathrm{Re}\{\sqrt{\cdot}\} > 0$. Let us write the zero term separately:

$$\sum_{n=-\infty}^{\infty} e^{-jqnd} H_0^{(2)} \left(k\sqrt{z^2 + (nd - y)^2} \right) = \frac{2}{d} \frac{1}{\sqrt{k^2 - q^2}} e^{-j\sqrt{k^2 - q^2}|z|}$$

$$+\frac{4j}{d}\sum_{m=1}^{\infty}\cos\left(\frac{2\pi my}{d}\right)\frac{e^{-\alpha_{mz}|z|}}{\alpha_{mz}} \qquad (4.15)$$

Finally, the field created by the array reads

$$E_x = -\frac{\eta}{2d}\frac{1}{\sqrt{1-q^2/k^2}}e^{-j\sqrt{k^2-q^2}|z|}I_0 - j\frac{\eta k}{d}\sum_{m=1}^{\infty}\cos\left(\frac{2\pi my}{d}\right)\frac{e^{-\alpha_{mz}|z|}}{\alpha_{mz}}I_0$$

$$(4.16)$$

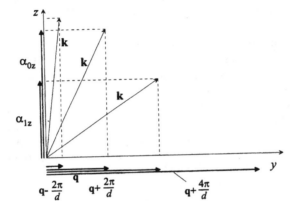

Figure 4.2 Propagation directions for the first several Floquet modes. In this example, three modes with the y-components of the wave vector equal to q and $q \pm 2\pi/d$ are propagating modes. The propagation directions are shown by vector **k**.

Each member of the series in (4.16) is a wave whose propagation factor along the grid plane is $q + 2\pi m/d$ [see (4.13)], so they are Floquet modes as defined by (4.4). This decomposition has a clear physical meaning. The first term is the plane wave excited by the averaged current of the grid I_0/d. It is a plane wave traveling from the array plane. If the grid period is small compared to the wavelength, all the other terms (*higher-order Floquet modes*) are exponentially decaying from the array plane. Note the imaginary unit before the sum in (4.16): These localized fields are *reactive*. Indeed, the coefficient that connects the electric field in the array plane E_x and the wire current I_0 has the meaning of the impedance per unit length of a wire in the grid. In this case it is an imaginary number, corresponding to a reactance. Considering the first term in (4.16) (a traveling plane wave), we find that for the field at the array plane $z = 0$, the coefficient is real. This reflects the fact that the traveling plane wave described by this term transports power away from the grid plane: Each line of current radiates some power in free space.

Suppose now that the array period is small compared with the wavelength in space and with the wavelength of the wave traveling along the grid, or, in terms of wavenumbers, $kd \ll 1$ and $qd \ll 1$. In this situation, (4.14) for $m \neq 0$ simplifies:

$$\alpha_{mz} = \sqrt{\left(\frac{2\pi m}{d} + q\right)^2 - k^2} \approx \frac{2\pi m}{d} \qquad (4.17)$$

($2\pi m/d \neq q$ because $qd \ll 1$). From (4.16) we see that the reactive field of dense arrays given by the sum of the higher-order Floquet modes becomes very small already at distances of the order of the grid period d, and the field of the grid is given by the plane wave excited by the averaged current. We can say that from distances larger than d the individual wires are not distinguishable by a probe antenna. The Floquet representation is very convenient for calculation of the field at far enough distances from the array plane, but for the observation points just at the array plane it cannot be used, because the series in (4.16) does not converge.[1]

With increasing frequency, the wavenumber k in formula (4.14) increases, the reactive fields decay more slowly, and eventually α_{mz} becomes imaginary, at first for $m = 1$. If $q = 0$, this happens at $kd = 2\pi$ or $d = \lambda$. Now, the grid creates *two* (or more, at still larger frequencies) traveling plane waves. This is illustrated in Figure 4.2, which shows the components of the wave vectors for the first Floquet modes. In the case shown in this figure, modes starting from $m = 2$ do not propagate, because $k < q + 4\pi/d$, and form reactive fields near the grid plane. The three first modes propagate. In the theory of phased array antennas [4] these traveling modes are called *grating lobes*.

As we will see, near the frequencies where new traveling waves appear, the reflection and transmission coefficients for grids have peculiarities called *Wood anomalies*.[2]

4.2 AVERAGED BOUNDARY CONDITION FOR A SIMPLE STRUCTURE: ARRAY OF WIRES

Electromagnetic properties of wire grids have been studied for many decades, because of their many important technical applications. Dense meshes of conducting wires (used as, e.g., screens or antenna reflectors) contain so many individual wires that any exact solution for nontrivial cases is not feasible, even using numerical techniques. Here we will show how wire meshes can be modeled by transition boundary conditions. This model essentially means that the actual complicated current structure is replaced by an equivalent smooth current distribution that creates nearly the same field at some distance from the array surface. Clearly, this is a good approximation for dense

---

[1]Naturally, an electric field in the array plain exists, but this method to compute it is not applicable.

[2]There exist other types of Wood anomalies that are not considered here.

grids in terms of the wavelength, as we have just seen from the properties of Floquet modes. In fact, introducing higher-order terms in the surface impedance allows even sparse grids to be modeled in this way. The modeling method makes use of the fact that for a simple planar regular grid of thin wires excited by plane waves, the exact analytical solution exists (see Section 4.2.1). This allows us to solve for the Fourier-transformed currents in an infinite planar grid excited by a plane wave. Now suppose that the grid is excited by an arbitrary source, but in such a way that the exciting field varies slowly along the array surface. "Slowly" means that it does not substantially change at distances of order d. Also, suppose that the grid is not infinite, but its size is large compared to the wavelength. Finally, the grid surface can be curved, but the model of smooth current is applicable only if the curvature radius is much larger than the wavelength. This is the typical situation in many applications for various screens and antenna reflectors.

Under the above assumptions, the current induced in a certain wire of the grid can be approximately determined from the knowledge of fields in the near vicinity of that wire. In other words, a local or higher-order impedance boundary condition can be established, similar to the case of thin material layers that we considered in Chapters 2 and 3. This gives a way to find analytical solutions for rather involved systems, including grid edge diffraction and curved grids. The main problem in this theory is to find the boundary conditions in the physical space, transforming the exact solution from the Fourier domain. In principle, the basic physical idea of this approach is the same as used in the modeling of thin material layers in Chapter 2.

Approximate second-order boundary conditions, so-called *averaged boundary conditions*, have been known for a long time as models for dense grids of conducting wires [5,6] and strip arrays [7,8]. These boundary conditions can be used under the assumption that the grid period is considerably smaller than the wavelength. In the theory of the averaged boundary conditions [5,6], the local field exciting a reference wire is approximately calculated using the Euler-MacLaurin summation formula. Here we use an alternative approach [9], starting from the exact solution of the corresponding diffraction problem in the Fourier-transformed domain (for plane waves), and then finding the approximate impedance boundary condition, making an appropriate Taylor expansion of the exact result before transforming the fields to the physical space. This way we establish a method that allows us to derive higher-order boundary conditions of any order. In the second order we will obtain, naturally, the classical result of Kontorovich [5,6]. This problem has a long history, and similar physical ideas can be applied to modeling arrays of strips, and so forth, (see [7, 10–19]). We will discuss various extensions further in this chapter.

4.2.1   Summation of Wire Fields

As was mentioned above, we will develop an approximate model making use of the exact solution of a simplified problem. To this end, we start from the classical problem of plane-wave diffraction by regular arrays of thin parallel conducting wires. For planar grids of parallel thin wires excited by plane waves, the exact solution for the wire currents can be easily written in terms of a series of Hankel functions. This series converges rather slowly, but the convergence can be improved using the Poisson summation rule and some other means, as shown in [15,16]. We have already made use of this technique in the discussion of Floquet modes.

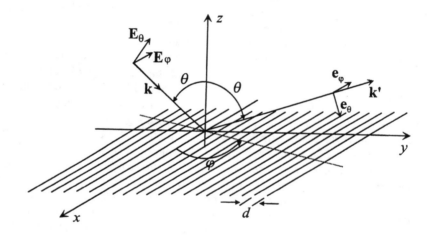

Figure 4.3 The geometry of the planar regular array of wires.

Consider a periodic grid of parallel conducting wires [9] (Figure 4.3). The wires can be ideally conducting or lossy wires whose impedance per unit length is Z. The wire radius is r_0, and the separation between the wires is d. In this theory, we assume that the wire radius is small compared to the wavelength ($kr_0 \ll 1$) and to the distance between the wires ($r_0 \ll d$). This means that we do not have to study the current density distribution over the wire cross section and can replace the wire field by the field of a line current I. The grid is excited by a plane wave

$$\mathbf{E}^{\text{ext}} = \mathbf{E}\, e^{-j\mathbf{k}\cdot\mathbf{r}} \tag{4.18}$$

so that the currents I_n in the wires should depend on the coordinates as

$$I_n = I\, e^{-jk_x x}\, e^{-jk_y y_n} \tag{4.19}$$

where $y_n = nd$ is the y-coordinate of the nth wire. The incidence and reflection angles are equal to θ (the normal incidence corresponds to $\theta = 0$), and the azimuthal incidence angle is φ. The Cartesian components of the wave vector $\mathbf{k}$ of the incident field can be written as:

$$k_x = k \sin \theta \cos \varphi, \qquad k_y = k \sin \theta \sin \varphi, \qquad k_z = -k \cos \theta \qquad (4.20)$$

The amplitude of the incident electric field $\mathbf{E}$ can be represented by its components $\mathbf{E}_\theta$ and $\mathbf{E}_\varphi$ connected with the Cartesian components as follows (see Figure 4.3):

$$\mathbf{E}_\theta = E_x \cos \theta \cos \varphi \, \mathbf{x}_0 + E_y \cos \theta \sin \varphi \, \mathbf{y}_0 + E_z \sin \theta \, \mathbf{z}_0$$

$$\mathbf{E}_\varphi = -E_x \sin \varphi \, \mathbf{x}_0 + E_y \cos \varphi \, \mathbf{y}_0 \qquad (4.21)$$

Each wire (marked by its number n) creates an electric field whose component in the direction along the wires (x-direction) is [3, p. 492]

$$E_{xn}^{\mathrm{w}} = -\frac{\eta}{4k}(k^2 - k_x^2)I_n H_0^{(2)}\left(\sqrt{k^2 - k_x^2}\, r_n\right) \qquad (4.22)$$

As before, $r_n = \sqrt{y_n^2 + z^2}$, $k = \omega\sqrt{\epsilon_0\mu_0}$ is the wave number, $\eta = \sqrt{\mu_0/\epsilon_0}$ is the wave impedance, and $H_0^{(2)}$ is the Hankel function. We used this result [equation (4.9)] for a simpler case of $k_x = 0$ when we considered Floquet modes in this system. To solve for the wire currents, we need to know the local field which acts on the wire surfaces. For the reference wire at $y = 0$ this field is expressed as

$$E_x^{\mathrm{loc}} = E_x e^{-jk_x x} - \frac{\eta}{2k}(k^2 - k_x^2)I e^{-jk_x x} \sum_{n=1}^{\infty} \cos(k_y nd) H_0^{(2)}\left(\sqrt{k^2 - k_x^2}\, nd\right) \qquad (4.23)$$

where the first term is the incident field and the sum gives the interaction field created by the array itself. The sum of the Hankel functions can be calculated using [20, formulas 8.521.1 and 8.522.3]:

$$\sum_{n=1}^{\infty} \cos(k_y nd) H_0^{(2)}\left(\sqrt{k^2 - k_x^2}\, nd\right) = \frac{1}{|k_z|d} - \frac{1}{2} +$$

$$\frac{j}{\pi}\left[\log\frac{\sqrt{k^2 - k_x^2}\, d}{4\pi} + \gamma + \frac{1}{2}\sum_{n=-\infty}^{\infty}{}' \left(\frac{2\pi}{\sqrt{(2\pi n + k_y d)^2 - (k^2 - k_x^2)d^2}} - \frac{1}{|n|}\right)\right] \qquad (4.24)$$

Here, $\gamma \approx 0.5772$ is the Euler constant [20, formula 9.73]. The prime in the sum denotes that the summation is made over all n except $n = 0$. The use of (4.23) and (4.24) leads to the exact (valid for thin wires) solution of the plane wave diffraction problem, (see [15, 16]). Remember that the Floquet expansion (in other words, the Poisson summation rule) leading to (4.16) is useless in this case since we are interested in the fields just at the array plane.

Now we can write the boundary condition on the surface of one of the wires (wire $n = 0$):

$$E_x^{\text{loc}} + E_{x0}^{\text{w}} = ZIe^{-jk_x x} \tag{4.25}$$

Here, E_{x0}^{w} is the field created by the reference wire current at that wire surface:

$$E_{x0}^{\text{w}} = -\frac{\eta}{4k}(k^2 - k_x^2)Ie^{-jk_x x}H_0^{(2)}\left(\sqrt{k^2 - k_x^2}\, r_0\right) \tag{4.26}$$

For ideally conducting wires, $Z = 0$ and this boundary condition simply means that the tangential electric field should vanish on the wire surface. For $Z \neq 0$, this is the relation between the wire current I and the voltage per unit length measured along the wire.

Under the assumption that $kr_0 \ll 1$, we replace the Hankel function by its asymptotic expression for small arguments,

$$H_0^{(2)}\left(\sqrt{k^2 - k_x^2}\, r_0\right) \approx 1 - j\frac{2}{\pi}\left(\log\frac{\sqrt{k^2 - k_x^2}\, r_0}{2} + \gamma\right) \tag{4.27}$$

which leads to:

$$E_x - \frac{\eta}{2k}(k^2 - k_x^2)I\left\{\frac{1}{|k_z|d} + \frac{j}{\pi}\left[\log\frac{d}{2\pi r_0}\right.\right.$$

$$\left.\left. + \frac{1}{2}\sum_{n=-\infty}^{\infty}{}'\left(\frac{2\pi}{\sqrt{(2\pi n + k_y d)^2 - (k^2 - k_x^2)d^2}} - \frac{1}{|n|}\right)\right]\right\} = ZI \tag{4.28}$$

It is important to note that the real part of the reference wire field [it comes from the real part of the Hankel function in (4.27)] cancels out with the term $-1/2$ in summation (4.24). This fact has an important physical meaning. Indeed, the real part of the reference wire field refers to its radiation resistance and describes the scattering by an individual wire. In dense regular arrays there is no scattering: If an array is excited by a plane wave, energy is transported only by one reflected and one transmitted plane wave. This cancellation means that the interaction field induced by the other wires exactly compensates the scattered field of the reference wire. In the final expression for the *total* field in the array plane of dense grids, the real part contains only one term, which comes from $1/|k_z|d$ in (4.24). This is the plane-wave field created by the averaged current I/d. In the boundary condition to be introduced in the next section, this is reflected in the fact that the grid impedance is purely imaginary for lossless wires. These considerations are also important in understanding of composite materials, and more on this topic will be discussed in Section 5.2.2. When the distance between wires is larger than one half of the wavelength, one or more real terms emerge from the summation in (4.24), corresponding to grating lobes that take energy away from the incident plane wave and transform that into other propagating modes.

Since we want to establish a transition condition for cell-averaged quantities, it is convenient to introduce the cell-averaged surface current density $\widehat{J}$ connected to the current I as $\widehat{J} = I/d$. From (4.28) we determine the wire current and the averaged current density:

$$\widehat{J} = \frac{2}{\eta} \frac{\frac{|k_z|}{k}}{\left(1 - \frac{k_x^2}{k^2}\right)\left(1 + j\alpha\frac{|k_z|}{k}\right) + \frac{2}{\eta}\frac{|k_z|}{k}Zd} E_x \qquad (4.29)$$

or, in terms of the incidence angles,

$$\widehat{J} = \frac{2}{\eta} \frac{\cos\theta}{(1 - \sin^2\theta\cos^2\varphi)(1 + j\alpha\cos\theta) + (2/\eta)Zd\cos\theta} E_x \qquad (4.30)$$

Here we have introduced an important parameter

$$\alpha = \frac{kd}{\pi}\left[\log\frac{d}{2\pi r_0} + \frac{1}{2}\sum_{n=-\infty}^{\infty}{}'\left(\frac{2\pi}{\sqrt{(2\pi n + k_y d)^2 - (k^2 - k_x^2)d^2}} - \frac{1}{|n|}\right)\right] \qquad (4.31)$$

which is called *grid parameter*, because its value determines the grid properties. The series in (4.31) converges very quickly. For dense grids, such that $k_y d \ll 2\pi$ and $\sqrt{k^2 - k_x^2}\, d \ll 2\pi$, it gives a very small correction to the first logarithmic term and can be neglected. In that case, the averaged current density is given by (4.30) with α replaced by α_{ABC}:

$$\alpha_{\text{ABC}} = \frac{kd}{\pi}\log\frac{d}{2\pi r_0} \qquad (4.32)$$

In this notation, ABC stands for the averaged boundary condition, which we will introduce next.

4.2.2 Averaged Boundary Condition

To find a transition condition that connects cell-averaged tangential fields at the array plane, there is no need to perform spatial averaging. It is enough to notice that the averaged field equals the field created by the averaged surface current $\widehat{J}$. Thus, the x-component of the total (Fourier-domain) averaged electric field $\widehat{E}_x^{\text{tot}}$ in the grid plane is the sum of the incident field $E_x^{\text{ext}} = E_x\, e^{-jk_x x}\, e^{-jk_y y}$ and the x component of the plane-wave field created by the current sheet $\widehat{J} = J\, e^{-jk_x x}\, e^{-jk_y y}$. To find the x-component of the electric field created by this current sheet, we first note that the tangential component of the averaged scattered magnetic field reads

$$\mathbf{z}_0 \times \mathbf{H}_t = \frac{\widehat{J}}{2}\mathbf{x}_0 = \frac{\widehat{J}}{2}\left(\cos\varphi\frac{\mathbf{k}_t}{|k_t|} - \sin\varphi\frac{\mathbf{z}_0 \times \mathbf{k}_t}{|k_t|}\right) \qquad (4.33)$$

The tangential electric field can now be found from (2.15):

$$\mathbf{E}_t = -\frac{\eta \widehat{J}}{2} \left(\sqrt{1 - \frac{k_t^2}{k^2}} \cos \varphi \, \frac{\mathbf{k}_t}{|\mathbf{k}_t|} - \frac{\sin \varphi}{\sqrt{1 - \frac{k_t^2}{k^2}}} \, \frac{\mathbf{z}_0 \times \mathbf{k}_t}{|\mathbf{k}_t|} \right) \tag{4.34}$$

The x-component (along the wires) of the averaged electric field created by the wire grid is then

$$\widehat{E_x} = -\frac{\eta \widehat{J}}{2} \frac{\left(1 - \frac{k_t^2}{k^2}\right) \cos^2 \varphi + \sin^2 \varphi}{\sqrt{1 - \frac{k_t^2}{k^2}}} = -\frac{\eta}{2} \frac{1 - \sin^2 \theta \cos^2 \varphi}{\cos \theta} \, \widehat{J} \tag{4.35}$$

Summing up the incident and scattered fields, we get

$$\widehat{E}_x^{\text{tot}} = E_x^{\text{ext}} - \frac{\eta}{2} \frac{1 - \sin^2 \theta \cos^2 \varphi}{\cos \theta} \, \widehat{J} \tag{4.36}$$

Substitution of the current density from (4.30) gives for the averaged total field

$$\widehat{E}_x^{\text{tot}} = \frac{\left[Zd + j\frac{\eta}{2}\alpha(1 - \sin^2 \theta \cos^2 \varphi) \right] \frac{2}{\eta} \cos \theta}{(1 - \sin^2 \theta \cos^2 \varphi)(1 + j\alpha \cos \theta) + (2/\eta)Zd \cos \theta} E_x^{\text{ext}} \tag{4.37}$$

Expressing the same result in terms of the averaged current $\widehat{J}$, we get

$$\widehat{E}_x^{\text{tot}} = \left[Zd + j\frac{\eta}{2}\alpha(1 - \sin^2 \theta \cos^2 \varphi) \right] \widehat{J} \tag{4.38}$$

From (4.20), we recognize that $1 - \sin^2 \theta \cos^2 \varphi = 1 - k_x^2/k^2$.

Relation (4.38) connects the averaged induced current with the total averaged electric field in the array plane, but only for plane wave excitation. For arbitrary fields in the physical domain, we should substitute $k_x^2 \to -\partial^2/\partial x^2$ and $k_y^2 \to -\partial^2/\partial y^2$. Obviously, to arrive at effective boundary conditions we should find a rational approximation for the grid parameter α (4.31). The situation is rather similar to the case of modeling thin layers and interfaces (Chapters 2 and 3). This can be done by expanding α in the Taylor series with respect to d/λ, if we are interested in modeling arrays for waves incident from directions close to the normal direction z. If we only want to keep the second-order spatial derivatives in our approximate boundary conditions, it is enough to keep only the zero-order term in the expansion of α. This means neglecting the sum in (4.31), because its expansion starts from second-order terms. Thus, the second-order boundary condition reads

$$\widehat{E}_x^{\text{tot}} = \left[Zd + j\frac{\eta}{2}\alpha_{\text{ABC}} \left(1 + \frac{1}{k^2} \frac{\partial^2}{\partial x^2} \right) \right] \widehat{J} = Z_g \widehat{J} \tag{4.39}$$

Here we have replaced $1 - \sin^2\theta\cos^2\varphi$ by $1 - k_x^2/k^2$ and made the inverse Fourier transform.

Differential operator Z_g can be considered as the grid impedance in the equivalent transmission-line model. This is the well-known result (*Kontorovich averaged boundary condition*) established by Kontorovich [5,6], who used a different derivation approach. As we see, the result of the exact summation (4.38) differs from the approximate Kontorovich boundary condition (4.39) by an additional member in the expression for the grid parameter. Rational approximations (such as the Taylor expansion) of this additional term lead to higher-order boundary conditions [9]. This approach is valid until there is only one propagating Floquet mode created by the wire grid currents. At higher frequencies, the very assumption that at a distance from the grid we observe only reflected and transmitted waves created by the averaged current is not correct, which means that we cannot write a relation between the averaged current and the averaged electric field as in (4.39).

4.2.3   Equivalent Transmission-Line Formulation

Identifying the electric field along the wires with the voltage of an equivalent transmission line and the averaged current density $\widehat{J}$ with its current, the reflection and transmission coefficients can be conveniently found from the transmission-line theory. Of course, the same results can be found directly from the averaged boundary condition, but the transmission-line formalism is sometimes more convenient and helps to get more physical insight. In this model, the grid is modeled by a shunt load in a transmission line. The load impedance equals the grid impedance Z_g (4.39). The characteristic impedance of the transmission line equals the wave impedance of the medium in which the grid is located. For the normal incidence of waves, this is simply the wave impedance $\eta = \sqrt{\mu/\epsilon}$ (see Figure 4.4).

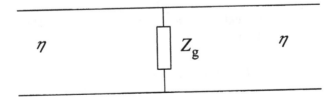

Figure 4.4 Equivalent transmission-line model for a planar array of wires in an isotropic space with the wave impedance η (normal incidence).

In terms of this model, the reflection and transmission coefficients can be derived in the following way. First, we write the boundary condition (4.39) as

$$\mathbf{E}^{\text{ext}} - \frac{\eta}{2}\widehat{\mathbf{J}} = Z_g\widehat{\mathbf{J}} \qquad (4.40)$$

(normal incidence is assumed). From here, the induced surface current density is found:

$$\widehat{\mathbf{J}} = \frac{\mathbf{E}^{\text{ext}}}{Z_g + \frac{\eta}{2}} \tag{4.41}$$

Finally, the reflected plane-wave field created by the current sheet $\widehat{\mathbf{J}}$ is

$$\mathbf{E}^{\text{ref}} = -\frac{\eta}{2}\widehat{\mathbf{J}} \tag{4.42}$$

thus, the reflection and transmission coefficients read

$$R = -\frac{\frac{\eta}{2}}{Z_g + \frac{\eta}{2}}, \qquad T = 1 + R = \frac{Z_g}{Z_g + \frac{\eta}{2}} \tag{4.43}$$

4.2.4   Another Form of the Averaged Boundary Condition

The boundary condition can be also written in terms of the tangential component of the averaged magnetic field on the two sides of the grid, because $\widehat{J} = \widehat{J}_x = \widehat{H}_{y-} - \widehat{H}_{y+}$, where the subscripts $\pm$ mark the averaged fields on the sides $z > 0$ and $z < 0$, respectively. Thus, we get

$$\widehat{E}_x^{\text{tot}} = \left[Zd + j\frac{\eta}{2}\alpha_{\text{ABC}} \left(1 + \frac{1}{k^2}\frac{\partial^2}{\partial x^2}\right) \right] (\widehat{H}_{y-} - \widehat{H}_{y+}) \tag{4.44}$$

Sometimes it can be better to write this condition in another form involving the normal component of the electric field in the array plane. Using the Maxwell equation

$$j\omega\epsilon E_z = \frac{\partial H_y}{\partial x} - \frac{\partial H_x}{\partial y} \tag{4.45}$$

and taking into account that there is no induced current in the y direction if the wires are very thin,[3] we replace

$$\frac{\partial}{\partial x}(\widehat{H}_{y-} - \widehat{H}_{y+}) \rightarrow j\omega\epsilon(\widehat{E}_{z-} - \widehat{E}_{z+}) \tag{4.46}$$

The result reads

$$\widehat{E}_x^{\text{tot}} = \left(Zd + j\frac{\eta}{2}\alpha_{\text{ABC}} \right)(\widehat{H}_{y-} - \widehat{H}_{y+}) - \frac{\alpha_{\text{ABC}}}{2k}\frac{\partial}{\partial x}(\widehat{E}_{z-} - \widehat{E}_{z+}) \tag{4.47}$$

Averaged boundary conditions in the last form can be found, for example, in [7, 19] (for arrays of ideally conducting strips).

---

[3]This means that H_x and $\partial H_x/\partial y$ are continuous across the array plane.

4.3 REFLECTION AND TRANSMISSION PROPERTIES OF WIRE GRIDS

The effective boundary conditions are primarily needed to solve diffraction problems for nonplane-wave excitation, or for limited in size or curved grids. Here, in order to check the accuracy of the solution, we will calculate exact and approximate reflection coefficients for infinite grids excited by plane waves. For this goal we find the reflection and transmission coefficients in terms of the grid parameter α. For the normal incidence, we have already written them in terms of the grid impedance in (4.43).

4.3.1 Reflection and Transmission Coefficients

In the far zone, the grid field is a plane-wave field with the components:

$$E_\theta^{\text{ref}} = \mathbf{e}_\theta \cdot \mathbf{E}^{\text{ref}} = -\frac{\eta}{2} \cos\varphi \, \widehat{J}, \qquad E_\varphi^{\text{ref}} = \mathbf{e}_\varphi \cdot \mathbf{E}^{\text{ref}} = \frac{\eta}{2} \frac{\sin\varphi}{\cos\theta} \, \widehat{J} \qquad (4.48)$$

where $\mathbf{e}_\theta$ and $\mathbf{e}_\varphi$ are the unit vectors connected with the unit vectors of the Cartesian coordinate system as follows (see Figure 4.3):

$$\mathbf{e}_\theta = \cos\theta\cos\varphi \, \mathbf{x}_0 + \cos\theta\sin\varphi \, \mathbf{y}_0 - \sin\theta \, \mathbf{z}_0, \qquad \mathbf{e}_\varphi = -\sin\varphi \, \mathbf{x}_0 + \cos\varphi \, \mathbf{y}_0 \qquad (4.49)$$

After substitution of the current (4.30), the co- and cross-polarized reflection coefficients are found as

$$R_{\text{TM}} = \frac{E_\theta^{\text{ref}}}{E_\theta} = -\frac{\cos^2\theta\cos^2\varphi}{(1 - \sin^2\theta\cos^2\varphi)(1 + j\alpha\cos\theta) + (2/\eta)Zd\cos\theta}$$

$$R_{\text{EM}} = \frac{E_\varphi^{\text{ref}}}{E_\theta} = \frac{\cos\theta\sin\varphi\cos\varphi}{(1 - \sin^2\theta\cos^2\varphi)(1 + j\alpha\cos\theta) + (2/\eta)Zd\cos\theta}$$

$$(4.50)$$

for TM-polarized incident waves, and

$$R_{\text{TE}} = \frac{E_\varphi^{\text{ref}}}{E_\varphi} = -\frac{\sin^2\varphi}{(1 - \sin^2\theta\cos^2\varphi)(1 + j\alpha\cos\theta) + (2/\eta)Zd\cos\theta}$$

$$R_{\text{ME}} = \frac{E_\theta^{\text{ref}}}{E_\varphi} = \frac{\cos\theta\sin\varphi\cos\varphi}{(1 - \sin^2\theta\cos^2\varphi)(1 + j\alpha\cos\theta) + (2/\eta)Zd\cos\theta}$$

$$(4.51)$$

for TE-polarized incident waves. Because the transmitted fields are the sums of the incident fields and the fields created by the grid,[4] the transmission coefficients are determined as

$$T_{\text{TM}} = 1 + R_{\text{TM}}, \qquad T_{\text{EM}} = R_{\text{EM}} \qquad (4.52)$$

---

[4] The amplitudes of waves generated by the grid are the same on both sides of the grid plane.

$$T_{\text{TE}} = 1 + R_{\text{TE}}, \qquad T_{\text{ME}} = R_{\text{ME}} \tag{4.53}$$

This follows from the continuity of the tangential electric field at the array plane: The wire grid supports only electric current.

4.3.2   Numerical Examples

In the following examples, some typical curves for the reflection coefficient from dense wire grids as functions of the incidence angle are shown. We observe that the model of approximate boundary conditions works very well for dense grids. In the examples given in Figure 4.5, the ratio between the wire radius and the array period is $r_0/d = 1/50$. The wires are ideally conducting. The two sets of curves correspond to $d/\lambda = 0.1$ (the grid parameter $\alpha = 0.41$) and $d/\lambda = 0.2$ (the grid parameter $\alpha = 0.83$). We can conclude that the averaged boundary condition for wire grids (4.39) provides a very accurate model for grids with the grid parameter $\alpha \leq 0.5$.

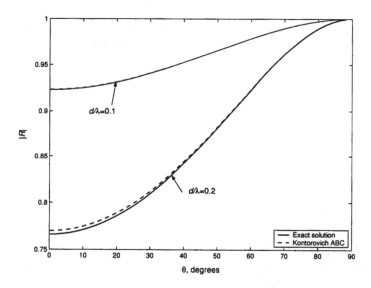

Figure 4.5 Angular dependence of the absolute value of the reflection coefficient from a dense grid of parallel conducting wires. The azimuthal incidence angle $\varphi = 0$. Comparison is between exact and approximate results.

The typical frequency behavior of the reflection and transmission coefficients is shown in Figure 4.6. In this example, the ratio of the wire radius r_0 to the grid period d equals 0.01. At low frequencies the grid is electrically dense and very reflective. There are resonance frequencies of the interaction field (at $d = n\lambda$, $n = 1, 2, 3, \dots$), where the reflection coefficient is zero and the transmission coefficient is one. The physical reason for this behavior can

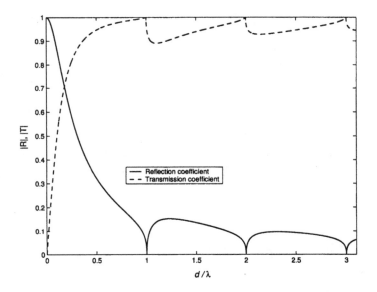

Figure 4.6 Absolute values of the reflection and transmission coefficients for a planar grid of parallel ideally conducting wires. Normal plane wave incidence.

be understood considering the interaction field induced by the whole grid at the surface of the reference wire. This field is, of course, proportional to the wire current. The coefficient (*interaction constant*) tends to infinity at these resonance frequencies because when the distance between wires is $n\lambda$, the contributions from all the wires are in phase. However, the field at the wire surface must be limited, which is only possible if the wire current is zero. Thus, at these resonant frequencies the grid cannot be excited, and the wave passes through without any reflection.

Looking at Figure 4.6, it is easy to notice that the usual energy conservation relation for lossless layers, $|R|^2 + |T|^2 = 1$, is not satisfied at high enough frequencies. This is illustrated by Figure 4.7, where this quantity is plotted against the normalized grid period for the same grid as in Figure 4.6. This is because at high frequencies some part of the incident plane wave power is taken away from the grid by grating (diffraction) lobes.

4.4 WIRE GRIDS: GENERALIZATIONS

4.4.1 Rectangular Cells

For dense grids with rectangular cells (cell size $a \times b$, b along the axis x and a along the axis y) made of ideally conducting wires with ideal contacts in

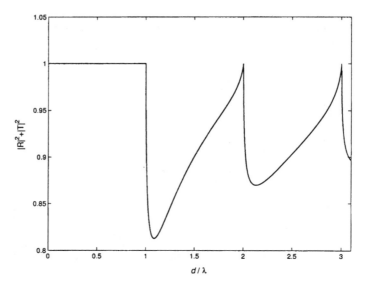

Figure 4.7 On the discussion of the energy conservation in reflection and transmission through regular grids.

the wire crossings, the averaged boundary conditions read [5]

$$E_x^{\text{tot}} = j\eta \frac{ka}{2\pi} \log \frac{a}{2\pi r_0} \left[\widehat{J}_x + \frac{1}{k^2(1 + b/a)} \frac{\partial}{\partial x} \left(\frac{b}{a} \nabla \cdot \widehat{\mathbf{J}} \right) \right] \tag{4.54}$$

$$E_y^{\text{tot}} = j\eta \frac{kb}{2\pi} \log \frac{b}{2\pi r_0} \left[\widehat{J}_y + \frac{1}{k^2(1 + a/b)} \frac{\partial}{\partial y} \left(\frac{a}{b} \nabla \cdot \widehat{\mathbf{J}} \right) \right] \tag{4.55}$$

Even more general cases of nonideally conducting wire meshes with arbitrary contacts at the crossing points were also studied in [5].

4.4.2   Grids of Metal Strips

Next we will show how the averaged boundary conditions can be extended to other geometries of the unit cell, for example for grids of planar conducting strips (Figure 4.8).

As is obvious from the averaged boundary conditions for dense wire grids, their reflection and transmission properties are described by a single scalar parameter (*grid parameter*) α.[5] If for example the grid period changes, it is enough to change parameter α in the model. The analytical form of the boundary condition which determines how the reflection coefficient depends on the incidence angle is universal for dense arrays. A very illuminating

---

[5]Or two such parameters, as in (4.54) and (4.55), if the grid is anisotropic.

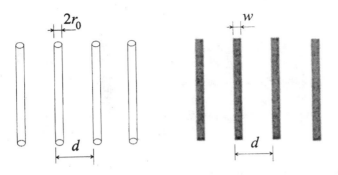

Figure 4.8 Grids formed of wires with round cross section and of planar conducting strips of negligible thickness and width w.

discussion of this feature can be found in [14], where it was established that one can measure the reflection and transmission coefficients from a grid at only one incidence angle and determine the array properties for all incidence angles. For our present goal this means that it is possible to determine the grid parameter, for example, for a strip in a waveguide (reflections in the walls simulate the infinite periodical array). This was done in [8]. The grid parameter for dense strip grids looks similar to that of a wire grid except for a different argument of the logarithmic factor:

$$\alpha_{\text{ABC}} = \frac{kd}{\pi} \log \frac{1}{\sin \frac{\pi w}{2d}} \tag{4.56}$$

For thin strips the argument of the sine function is small, and expanding it into the Taylor series we find approximately

$$\alpha_{\text{ABC}} = \frac{kd}{\pi} \log \frac{2d}{\pi w} \tag{4.57}$$

Comparing with α_{ABC} for grids of round wires (4.32), we see that with regard to the grid properties, a round wire of radius r_0 is equivalent to a strip of width $w = 4r_0$.

Arrays of round-cross-section conductors are usually made of thin wires, which is why we assumed that the wire polarization by electric fields orthogonal to the wires can be neglected. In some applications, arrays of strips have, on the contrary, very thin open gaps between the strips. Also, models for arrays with square cells like those shown in Figure 4.9 are needed. To extend the average boundary condition to cover these situations, we make use of the Babinet principle (it is enough to determine the grid parameter at one incidence angle, and we apply the Babinet principle for grids illuminated by normally incident plane waves).

Let the grid impedance of a strip grid (Figure 4.8, right side) for excitation by a plane wave with electric field polarized along the strips be Z_g. For

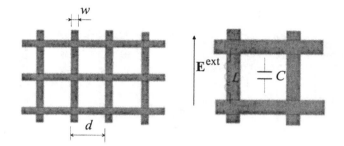

Figure 4.9 Grid of conducting strips with square cells.

the normal incidence of waves on a dense array of ideally conducting strips, the grid impedance

$$Z_g = j\frac{\eta}{2}\alpha_{\text{ABC}} = j\omega\mu_0\frac{d}{2\pi}\log\frac{1}{\sin\frac{\pi w}{2d}} \qquad (4.58)$$

is inductive [see (4.39)]. From the Babinet principle, we find that for excitation by a plane wave polarized orthogonally to the strip

$$Z_{g\perp} = \frac{\eta^2}{4Z_g} \qquad (4.59)$$

where η is the wave impedance of the surrounding medium, and the strip width w should be replaced by the gap width $d - w$ in the expression for Z_g. Substituting Z_g, we get

$$Z_{g\perp} = \frac{\pi}{j\omega\epsilon_0 2d\log\frac{1}{\sin\frac{\pi(d-w)}{2d}}} \qquad (4.60)$$

Clearly, the inductive impedance of strip grids polarized along the strips is transformed into a capacitive inductance if the electric field is applied orthogonally to the strips. For meshes with square (or rectangular) cells, the equivalent circuit of a cell is a parallel connection of an inductance and a capacitance (Figure 4.9). At the resonance of the cell, the grid becomes transparent.

4.5 DENSE ARRAYS OF PARTICLES

Periodical arrays of small scatterers are models for very thin material layers, frequency selective surfaces, reflect arrays, and so forth. Here we will learn how to understand interaction of electromagnetic waves with such structures.

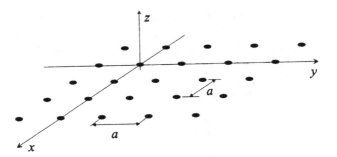

Figure 4.10 Planar regular array of small electrically or magnetically polarizable particles. The array extends to infinity along axes x and y.

4.5.1 The Local Field

Consider a two-dimensional periodical array of small identical particles (Figure 4.10). The assumption of smallness in this case means that it is enough to consider only electric or (and) magnetic dipoles induced in the particles and neglect the higher-order multipoles. In the most general case of bianisotropic inclusions, both electric and magnetic moments can be excited by both electric and magnetic fields [21], but for simplicity we assume that the particles can be polarized by either electric or magnetic field only. We will write formulas assuming electrically polarizable inclusions.

Let the grid have square cells of the size $a \times a$ and be excited by external electric field $\mathbf{E}^{\text{ext}}$ that is uniform in the array plane. In this situation all the particles are equivalent, and the induced dipole moment is the same for all particles. If we want to know, say, the reflection coefficient, we have to calculate the dipole moment of the inclusions. Suppose we know the properties of one individual particle (in our case, its polarizability dyadic that connects the induced dipole moment and the field exciting the particle). The key problem is the calculation of the field acting on every particle, the so-called *local field*, which is the sum of the external field $\mathbf{E}^{\text{ext}}$ and the field induced by all other inclusions $\mathbf{E}^{\text{int}}$ [22]. The last field is called the *interaction field* [23].

We can try to calculate the interaction field numerically, summing up the fields of the dipoles in the array. An appropriate limit should be taken because the grid contains infinitely many particles. To this end, let us introduce a circle of radius R around one of the particles and calculate

$$\mathbf{E}^{\text{hole}} = \sum \mathbf{E}_p \qquad (4.61)$$

summing up the fields created at the origin by all the dipoles sitting inside the circle of radius R except the reference dipole located at the origin. Next, we want to increase the radius R until the result converges. However, no limit exists if there are no losses in space. Indeed, the wave field of dipoles at distance R behaves as $1/R$, but the number of dipoles located at distance R increases proportionally to R, that is, to the perimeter of the circle. The result is an oscillating function of R. If we assume small losses in space, the limit does exist, but the series converges very slowly, and anyway it is not possible to take the limit to the lossless case.

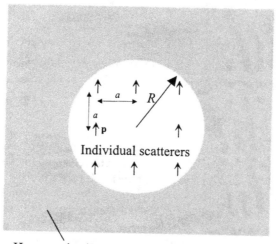

Homogenized current sheet

Figure 4.11 Calculation of the local field: individual dipoles near the reference particle and a homogenized current sheet far from the reference particle. Electric dipole moments of small particles are shown by the small arrows. The reference dipole at the center has been removed.

To overcome this difficulty, we introduce small losses in space, assuming that the wave number k has a small negative imaginary part, and split the interaction field into two parts: the field of the dipoles inside the circle of radius R and the field of the dipoles located outside this circle. The calculation of the second part can be simplified if we assume that we can replace the discrete set of dipoles by an averaged continuous distribution of polarization, which makes sense for large R. This is because from a large enough distance, the reference dipole cannot "see" the individual dipoles but rather feels the field of the averaged polarization. Thus, we write

$$\mathbf{E}^{loc} = \mathbf{E}^{ext} + \mathbf{E}^{sheet} + \mathbf{E}^{hole} \tag{4.62}$$

where $\mathbf{E}^{sheet}$ is the field of a current sheet with a hole of radius R, and $\mathbf{E}^{hole} = \sum \mathbf{E}_p$ is the sum of the fields of discrete dipoles within the circle of radius R. The sum of the two last members in (4.62) gives the interaction field:

$$\mathbf{E}^{\text{int}} = \mathbf{E}^{\text{sheet}} + \mathbf{E}^{\text{hole}} \tag{4.63}$$

This approach is illustrated by Figure 4.11. The field $\mathbf{E}^{\text{sheet}}$ can be calculated analytically. Clearly, the problem of the sum convergence is then solved because $\mathbf{E}^{\text{hole}}$ is the sum of the fields created by a finite number of dipoles (in practice, as we will see, by a very small number of dipoles).

The electric field generated by an electric dipole is (e.g., [24, p. 411])

$$\mathbf{E}_p = \frac{1}{4\pi\epsilon_0} \left\{ k^2 (\mathbf{n} \times \mathbf{p}) \times \mathbf{n} \frac{e^{-jkr}}{r} + [3\mathbf{n}(\mathbf{n} \cdot \mathbf{p}) - \mathbf{p}] \left(\frac{1}{r^3} + \frac{jk}{r^2} \right) e^{-jkr} \right\} \tag{4.64}$$

where $\mathbf{p}$ is the dipole moment, $\mathbf{n}$ is the unit vector directed from the source to the observation point, r is the distance from the source to the observation point, and $k = \omega\sqrt{\epsilon_0\mu_0}$. Here we have made no simplifications, and have included all the terms of the dipole field: the near-field contribution (quasistatic field), which is proportional to $1/r^3$, the intermediate-zone field $1/r^2$, and the wave field $1/r$. Magnetic field is not considered since the interaction magnetic field is zero (due to the symmetry of the problem).

To calculate $\mathbf{E}^{\text{sheet}}$ we first replace the discrete dipoles by a homogeneous polarized sheet. The dipole moment per unit area (the surface density of dipole moments) is

$$\mathbf{p}_s = \frac{\mathbf{p}}{a^2} \tag{4.65}$$

(a^2 is the cell area). Let us split vector $\mathbf{p}_s$ into two components, parallel and orthogonal to the array plane, and consider them separately. For the dipoles parallel to the plane, we can introduce an equivalent averaged electric current surface density

$$\widehat{\mathbf{J}} = j\omega\mathbf{p}_s \tag{4.66}$$

Assuming that there are (infinitesimally small) losses in space ($k = k' - jk''$, $k'' > 0$), the field created by the current sheet with a hole of radius R can be calculated as

$$\mathbf{E}^{\text{sheet}} = \frac{1}{4\pi\epsilon_0} \int \left\{ k^2 (\mathbf{n} \times \mathbf{p}_s) \times \mathbf{n} + [3\mathbf{n}(\mathbf{n} \cdot \mathbf{p}_s) - \mathbf{p}_s] \frac{1 + jkr}{r^2} \right\} \frac{e^{-jkr}}{r} \, dS \tag{4.67}$$

where the surface integral is taken over the area defined by $r > R$. Next we introduce a polar coordinate system. For the component of $\mathbf{p}_s$ parallel to the plane, the angle φ is measured from the direction of $\mathbf{p}_s$. In this case, $\mathbf{E}^{\text{sheet}}$ has only one nonzero component (parallel to $\mathbf{p}_s$). In the polar coordinates

$$\mathbf{E}^{\text{sheet}} = \frac{\mathbf{p}_s}{4\pi\epsilon_0} \int_0^{2\pi} \int_R^{\infty} \left\{ k^2 \sin^2\varphi + (3\cos^2\varphi - 1)\frac{1 + jkr}{r^2} \right\} e^{-jkr} \, dr \, d\varphi \tag{4.68}$$

This integral can be expressed in closed form [22]. The result reads

$$\mathbf{E}^{\text{sheet}} = -j\omega \mathbf{p}_s \frac{\eta}{4}\left(1 - \frac{1}{jkR}\right)e^{-jkR} = \hat{\mathbf{J}}\frac{\eta}{4}\left(1 - \frac{1}{jkR}\right)e^{-jkR} \qquad (4.69)$$

We have used notation (4.66) (and, as before, $\eta = \sqrt{\mu_0/\epsilon_0}$). For dipoles orthogonal to the array plane, we get in the same way[6]

$$\mathbf{E}^{\text{sheet}} = \frac{\mathbf{p}_s}{4\pi\epsilon_0}\int_0^{2\pi}\int_R^{\infty}\left\{k^2 - \frac{1+jkr}{r^2}\right\}e^{-jkr}\,dr\,d\varphi \qquad (4.70)$$

After integration, (4.70) reduces to

$$\mathbf{E}^{\text{sheet}} = -j\omega\mathbf{p}_s\frac{\eta}{2}\left(1 + \frac{1}{jkR}\right)e^{-jkR} \qquad (4.71)$$

Now that the integrals have been calculated, we can take the limit $\text{Im}\{k\} \to 0$ and use these results for lossless matrices.

Let us analyze expressions (4.69) and (4.71) for $R \to 0$ and $R \to \infty$. For small kR we can use the Taylor expansion of the exponents in (4.69) and (4.71). In this way, we get

$$\mathbf{E}^{\text{sheet}} \approx \frac{\eta}{4jkR}\hat{\mathbf{J}} - \frac{\eta}{2}\hat{\mathbf{J}} + O(kR) \qquad (4.72)$$

for dipoles parallel to the plane, and

$$\mathbf{E}^{\text{sheet}} \approx -\omega\mathbf{p}_s\frac{\eta}{2kR} + O(kR) = -\frac{\mathbf{p}_s}{\epsilon_0 R} + O(kR) \qquad (4.73)$$

for dipoles orthogonal to the plane. In (4.72) we find the term $-\hat{\mathbf{J}}\eta/2$, which is the field of a plane wave excited by the complete (that is, without a hole) current sheet $\hat{\mathbf{J}}$. In case of the orthogonal orientation of dipoles, no plane wave is generated, and this term is absent in (4.73). We observe that with $kR \to 0$ the field $\mathbf{E}^{\text{sheet}}$ (its reactive part) tends to infinity. This is because the sources get closer and closer to the observation point. For $R \to \infty$ and lossy background media, $\mathbf{E}^{\text{sheet}}$ tends to zero in both cases. If there is no loss, we have in the limit an oscillating function (in the absolute value), and the amplitude of oscillations does not decrease with increasing R.

Now, to find the full interaction field, we should sum the fields of the dipoles within the circle R with the field $\mathbf{E}^{\text{sheet}}$ (4.69) or (4.71). As has been noted, both contributions behave similarly at large kR: as oscillating

---

[6]To make writing concise, we do not introduce special notations for the two components of the dipole moment density. In (4.69), $\mathbf{p}_s$ is the component parallel to the plane, but in (4.71), $\mathbf{p}_s$ is the component orthogonal to the array plane.

functions. However, the oscillations of the two parts of the interaction field are out of phase (for large enough kR), and they cancel out. This takes place starting from a certain (large enough) value of the hole radius R, which depends on the frequency and the grid period. In Figures 4.12 and 4.13, numerical results for $\mathbf{E}^{\text{hole}}$ and $\mathbf{E}^{\text{int}} = \mathbf{E}^{\text{hole}} + \mathbf{E}^{\text{sheet}}$ as functions of R/a are shown for two values of the normalized frequency: $ka = 0.13$ and $ka = 1.3$. In both cases the dipoles are parallel to the array plane. The fields are normalized and presented as the dimensionless *interaction constant* defined as in [23]: $C = E^{\text{int}}\epsilon_0 a^3/p$. The solid and dashed lines show the real and imaginary parts of the normalized field $\mathbf{E}^{\text{hole}}$, respectively. Symbols $\times$ and $\circ$ stand for the real and imaginary parts of the normalized total interaction field $\mathbf{E}^{\text{int}} = \mathbf{E}^{\text{hole}} + \mathbf{E}^{\text{sheet}}$, respectively. Analyzing the graphs, we see that with increasing R the sum of the dipole fields $\mathbf{E}^{\text{hole}}$ indeed does not tend to any limit but oscillates. We know that the field of the sheet with a hole behaves similarly. However, we observe that the total interaction field, calculated as the sum of $\mathbf{E}^{\text{hole}}$ and $\mathbf{E}^{\text{sheet}}$, quickly converges with increasing the hole radius R: For $ka = 0.13$ it is enough to take $R \approx 5a$, and for $ka = 1.3$, $R \approx 15a$. Similar conclusions can be made for dipoles oriented orthogonally to the array plane.

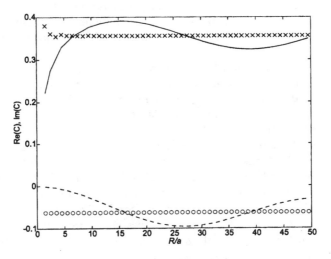

Figure 4.12 The sum of the dipole fields for $ka = 0.13$. Dipoles are parallel to the array plane. Solid and dashed lines show $\mathbf{E}^{\text{hole}}$ (real and imaginary parts, respectively), and symbols $\times$ and $\circ$ show the full interaction field $\mathbf{E}^{\text{int}} = \mathbf{E}^{\text{hole}} + \mathbf{E}^{\text{sheet}}$.

This brings us to a very important conclusion: The interaction field in infinite arrays is determined by a few nearest inclusions. This means that if the grid is not infinite or located on a curved surface, or the incident field is

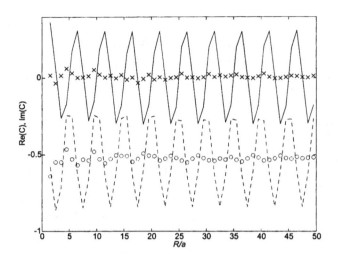

Figure 4.13 The same results as in Figure 4.12 for $ka = 1.3$.

not uniform, the estimation for the local field obtained for an infinite planar grid is still approximately valid. Obviously, this is true if the curvature radius is much larger than the grid period. Also, remember that inclusions located very near to the edge of the grid feel a (slightly) different local field than the inclusions inside the grid.

Solid lines in Figures 4.14 and 4.15 show the dependence of the interaction constant on $ka = 2\pi a/\lambda$. For dense grids when $ka \to 0$ the results tend to the static values [23, 25]. Dynamic results obtained by alternative methods are known for the parallel orientation of dipoles [23, p. 784], and the agreement is good up to $ka \approx 3$.

We see that for dense arrays only a few nearby particles should be considered separately. Consider the limiting case when the radius R is so small that there are actually no particles at all inside the circle (except one particle in the center); that is, $R < a$. In this situation, the full interaction field is given simply by the analytical expressions for $\mathbf{E}^{\text{sheet}}$ (4.69) and (4.71). It is easy to find such a value for R, which gives the correct static value for the interaction constant. The result is $R = R_0 = a/1.438$, which is the same for any orientation of the dipoles. Next, we can substitute this effective value $R = R_0$ in (4.69), (4.71) and check the validity of this model for different frequencies or array densities. The corresponding curves are shown as dashed lines in Figures 4.14 and 4.15. We conclude that for $ka < 1$ the interaction field can be calculated using simple analytical formulas (4.69) and (4.71) with a very good accuracy.

Finally, knowing the interaction field we can write down the local field

$$\mathbf{E}^{\text{loc}} = \mathbf{E}^{\text{ext}} + \mathbf{E}^{\text{int}} = \mathbf{E}^{\text{ext}} + \beta\mathbf{p} \tag{4.74}$$

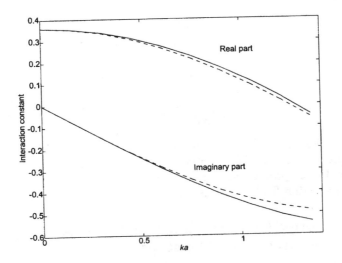

Figure 4.14 Dependence of the interaction constant C on ka. Dipoles are parallel to the array plane. Exact (solid curves) and approximate (dashed lines) results.

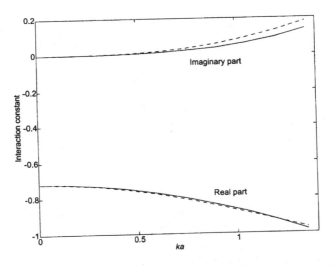

Figure 4.15 Dependence of the interaction constant C on ka. Dipoles are orthogonal to the array plane. Exact (solid curves) and approximate (dashed lines) results.

where $\mathbf{E}^{\text{ext}}$ is the external field, $\mathbf{p}$ is the dipole moment of one inclusion, and the (not normalized) *interaction constant* of the array β reads approximately

(4.69)

$$\beta \approx -j\frac{\omega}{a^2}\frac{\eta}{4}\left(1 - \frac{1}{jkR_0}\right)e^{-jkR_0} \qquad (4.75)$$

for the dipoles parallel to the array plane, and (4.71)

$$\beta \approx -j\frac{\omega}{a^2}\frac{\eta}{2}\left(1 + \frac{1}{jkR_0}\right)e^{-jkR_0} \qquad (4.76)$$

for the dipoles that are orthogonal to the array plane. With the knowledge of the interaction field, all electromagnetic properties of the array can be analyzed. Before proceeding in this direction, we will still improve the model for the interaction field using the energy conservation principle.

4.5.2 Energy Conservation and the Imaginary Part of the Interaction Constant

In the previous section, we found approximate analytical expressions for the real and imaginary parts of the interaction constant of dense arrays of dipole particles. Here we show that if the array is periodical, the imaginary part of the interaction constant can be found exactly from the energy conservation principle [26].

Let us first consider a single individual lossless dipole scatterer in free space. The induced dipole moment is proportional to the local field existing at the position of the particle:

$$p = \alpha E^{\text{loc}} \qquad (4.77)$$

where α is the particle polarizability. For simplicity we assume that the induced dipole moment is parallel to the applied field, so the polarizability α can be considered as a scalar. If we only have a single particle in free space, there is no interaction with other dipoles; hence, $E^{\text{loc}} = E^{\text{ext}}$ and the induced dipole moment is

$$p = \alpha E^{\text{ext}} \qquad (4.78)$$

The external field power spent on the excitation of the particle is proportional to the imaginary part of the polarizability α [27]:

$$P^{\text{ext}} = \frac{1}{2}\,\text{Re}\{J^* E^{\text{ext}}\} = \frac{1}{2}\,\text{Re}\{-j\omega p^* E^{\text{ext}}\}$$

$$= \frac{1}{2}\,\text{Re}\{-j\omega\alpha^*\}|E^{\text{ext}}|^2 = -\frac{\omega}{2}\,\text{Im}(\alpha)|E^{\text{ext}}|^2 \qquad (4.79)$$

(the asterisk $*$ denotes the complex conjugate operation). Similar expressions can be found elsewhere in the literature (e.g., [28]), but note that the present definition of the polarizability α includes also the effects of scattering, thus remaining a complex number even in the absence of absorption.

The total power radiated by the dipole is (e.g., [28])

$$P^{\text{rad}} = \frac{\mu_0 \omega^4 |p|^2}{12\pi c} = \frac{\mu_0 \omega^4 |\alpha|^2}{12\pi c} |E^{\text{ext}}|^2 \qquad (4.80)$$

where $c = 1/\sqrt{\epsilon_0 \mu_0}$ is the speed of light. If there is no absorption in the particle, all the received power is radiated back by the induced dipole: $P^{\text{rad}} = P^{\text{ext}}$, hence we have

$$\text{Im}\{\alpha\} = -\frac{\mu_0 \omega^3 |\alpha|^2}{6\pi c} \qquad (4.81)$$

or

$$\text{Im}\left\{\frac{1}{\alpha}\right\} = \frac{\eta \epsilon_0 \mu_0 \omega^3}{6\pi} = \frac{k^3}{6\pi \epsilon_0} \qquad (4.82)$$

Next, let us consider a regular array of lossless (ideally conducting) particles. Using the concept of the interaction constant (4.74) and equation (4.77), the dipole moment can be expressed in terms of the incident field as

$$p = \frac{\alpha}{1 - \alpha\beta} E^{\text{ext}} \qquad (4.83)$$

From this expression we find that the time-averaged power spent by the incident field on the excitation of one inclusion is, according to (4.79),

$$P^{\text{ext}} = -\frac{\omega}{2} \text{Im}\left\{\frac{\alpha}{1 - \alpha\beta}\right\} |E^{\text{ext}}|^2 \qquad (4.84)$$

On the other hand, in a regular dense array[7] there are no grating lobes and no scattering occurs: The whole array radiates only one plane wave, and every particle in the array radiates power

$$P^{\text{rad}} = 2a^2 \frac{1}{2} \text{Re}\{EH^*\} = a^2 \frac{\eta \widehat{J}}{2} \frac{\widehat{J}^*}{2} = \frac{\eta \omega^2}{4a^2} \left|\frac{\alpha}{1 - \alpha\beta}\right|^2 |E^{\text{ext}}|^2 \qquad (4.85)$$

In the first formula, a^2 is the cell area, and the coefficient 2 takes into account that the array radiates into two opposite directions from the array plane. The relation between the dipole moment of a single inclusion p and the averaged current density in the array plane $\widehat{J}$

$$\widehat{J} = \frac{j\omega p}{a^2} \qquad (4.86)$$

has been used. If there is no absorption in the particles, then $P^{\text{ext}} = P^{\text{rad}}$, so

$$\text{Im}\left\{\frac{\alpha}{1 - \alpha\beta}\right\} = -\frac{\eta \omega}{2a^2} \left|\frac{\alpha}{1 - \alpha\beta}\right|^2, \quad \text{or} \quad \text{Im}\left\{\frac{1}{\alpha} - \beta\right\} = \frac{\eta \omega}{2a^2} \qquad (4.87)$$

---

[7]The array period d is smaller than the wavelength λ.

Thus, we find that in the absence of absorption in particles, the imaginary part of the interaction constant is connected with the imaginary part of the particle polarizability as

$$\text{Im}\,\beta = \text{Im}\left\{\frac{1}{\alpha}\right\} - \frac{\eta\omega}{2a^2} \tag{4.88}$$

Now we can substitute $\text{Im}\{1/\alpha\}$ from (4.82) and determine the imaginary part of the interaction constant:

$$\text{Im}\,\beta = \frac{\eta\epsilon_0\mu_0\omega^3}{6\pi} - \frac{\eta\omega}{2a^2} = \frac{k^3}{6\pi\epsilon_0} - \frac{k}{2\epsilon_0 a^2} \tag{4.89}$$

The last result is valid not only for lossless particles because the interaction constant does not depend on whether the particles are lossy or lossless. Indeed, the value of the interaction constant is determined as a sum of dipole fields of an array of particles with the unit dipole moment. Equation (4.89) is an *exact* formula for the imaginary part of the interaction constant in arbitrary regular arrays of dipoles, provided that the distance between the particles is smaller than the wavelength.

The local field exciting every particle is given by (4.74) where β is

$$\beta = \text{Re}\left[-j\frac{\omega}{a^2}\frac{\eta}{4}\left(1 - \frac{1}{jkR_0}\right)e^{-jkR_0}\right] + j\left[\frac{\eta\epsilon_0\mu_0\omega^3}{6\pi} - \frac{\eta\omega}{2a^2}\right]$$

$$= \frac{\omega}{a^2}\frac{\eta}{4}\left(\frac{\cos kR_0}{kR_0} - \sin kR_0\right) + j\left[\frac{\eta\epsilon_0\mu_0\omega^3}{6\pi} - \frac{\eta\omega}{2a^2}\right] \tag{4.90}$$

Physically, this result means that the imaginary part of the interaction constant (which is responsible for the energy transport) is determined by the plane-wave field of the averaged current and the power scattered by one individual inclusion. The local field reads

$$E^{\text{loc}} = E^{\text{ext}} - \frac{\eta}{2}\widehat{J} + \frac{a^2\eta\epsilon_0\mu_0\omega^2}{6\pi}\widehat{J} - j\frac{\eta}{4}\left(\frac{\cos kR_0}{kR_0} - \sin kR_0\right)\widehat{J} \tag{4.91}$$

In the total averaged field (and, as we shall see, in the effective boundary condition) there is only the plane wave term, as should be, because the energy in regular arrays is transported by plane waves and no scattering occurs.

4.5.3 Equivalent Sheet Condition for Dense Arrays of Scatterers

Here we establish a boundary condition that connects cell-averaged electric and magnetic fields on the plane of dense arrays of dipole scatterers [26]. We consider regular arrays with square cells formed by small electric or magnetic dipole scatterers. To simplify the writing we consider only electric dipoles; corresponding results for magnetic dipoles follow from the duality principle. Each dipole particle can be lossy and is characterized by its (scalar) polarizability α. Here we study the case of the normal plane wave incidence. The incident electric field in the array plane is denoted as E^{ext}.

Boundary Condition

The total averaged (over one cell area) electric field in the plane of a dipole array $\widehat{E}^{\text{total}}$ is the sum of the incident field and the plane wave field created by the averaged current $\widehat{J}$:

$$\widehat{E}^{\text{total}} = E^{\text{ext}} - \frac{\eta}{2}\widehat{J} \tag{4.92}$$

In terms of the polarizability factor α and the interaction constant β, we can express the total field through the averaged current density only, because from (4.83) it follows that

$$E^{\text{ext}} = \frac{a^2}{j\omega}\left(\frac{1}{\alpha} - \beta\right)\widehat{J} \tag{4.93}$$

Inserting this expression into (4.92), we obtain

$$\widehat{E}^{\text{total}} = \left[\frac{a^2}{j\omega}\left(\frac{1}{\alpha} - \beta\right) - \frac{\eta}{2}\right]\widehat{J} = Z_g\widehat{J} \tag{4.94}$$

Using (4.90), this transforms to the final result

$$\widehat{E}^{\text{total}} = -j\frac{a^2}{\omega}\left\{\text{Re}\left(\frac{1}{\alpha} - \beta\right) + j\left[\text{Im}\left(\frac{1}{\alpha}\right) - \frac{k^3}{6\pi\epsilon_0}\right]\right\}\widehat{J} = Z_g\widehat{J} \tag{4.95}$$

This is the desired boundary condition since the current density is expressible in terms of the averaged tangential magnetic fields on the two sides of the array: $\widehat{J} = H_{t+} - H_{t-}$. Parameter Z_g is imaginary in the lossless case, because for nondissipating particles the imaginary part of the polarizability α satisfies (4.82), and the real part of the grid impedance vanishes. If there is some dissipation of energy in the particles, formula (4.95) is still valid. The imaginary part of the inverse polarizability $\text{Im}\{1/\alpha\}$ is in this situation larger than the scattering loss factor $k^3/(6\pi\epsilon_0)$. Although the scattering terms still cancel out, the grid impedance has a nonzero real part that describes absorption in the array.

In the low frequency limiting case ($ka \ll 1$), the boundary condition (4.95) is more convenient to write in terms of the particle dipole moment. In the quasistatic limit

$$\text{Re}\{\beta\} \approx \frac{0.36}{\epsilon_0 a^3} \tag{4.96}$$

(see Section 4.5.1, Figure 4.14). Using (4.86), we find that for arrays of lossless particles the result reads:

$$\widehat{E}^{\text{total}} = \left(\frac{1}{\alpha} - \frac{0.36}{\epsilon_0 a^3} \right) p \tag{4.97}$$

For example, for arrays of dielectric spheres of radius r and permittivity ϵ, substituting the sphere polarizability (e.g., [28])

$$\alpha = (\epsilon - \epsilon_0) \frac{3\epsilon_0}{\epsilon + 2\epsilon_0} V \tag{4.98}$$

where $V = 4\pi r^3/3$ is the sphere volume, we find

$$\widehat{E}^{\text{total}} = \frac{1}{\epsilon_0} \left(\frac{\epsilon + 2\epsilon_0}{\epsilon - \epsilon_0} \frac{1}{4\pi r^3} - \frac{0.36}{a^3} \right) p \tag{4.99}$$

Here the sphere permittivity ϵ can be complex to account for absorption losses.

Reflection and Transmission Coefficients

The boundary condition for regular dense arrays of dipole particles (4.95) connects the averaged electric and magnetic fields in the plane of the grid. Now it is very easy to solve for the reflection and transmission coefficients. From (4.92) and (4.95) the averaged surface current density can be solved:

$$\widehat{J} = \frac{1}{Z_g + \frac{\eta}{2}} E^{\text{ext}} \tag{4.100}$$

The reflected field is the plane wave created by this current; thus, the reflection coefficient is

$$R = -\frac{\eta}{2} \widehat{J} \frac{1}{E^{\text{ext}}} = -\frac{1}{1 + \frac{2Z_g}{\eta}} \tag{4.101}$$

The transmission coefficient

$$T = 1 + R = \frac{\frac{2Z_g}{\eta}}{1 + \frac{2Z_g}{\eta}} \tag{4.102}$$

For example, the reflection coefficient from an array of dielectric spheres in the low frequency limit reads

$$R = \frac{-ka}{ka - 2j \left(\frac{\epsilon + 2\epsilon_0}{\epsilon - \epsilon_0} \frac{a^3}{4\pi r^3} - 0.36 \right)} \tag{4.103}$$

The reflection and transmission of plane waves from dense arrays of dipole scatterers can be easily considered with the transmission-line model that we introduced as a model for wire and strip grids (Figure 4.4). Within the transmission-line model, Z_g plays the role of a shunt impedance simulating the grid properties (see Figure 4.4). For an array of dielectric spheres in the low frequency approximation, parameter Z_g is obtained from (4.99). Writing $Z_g = \frac{1}{j\omega C_s}$, the array is simply modeled by a shunt capacitance

$$C_s = \frac{\epsilon_0 a}{\frac{\epsilon + 2\epsilon_0}{\epsilon - \epsilon_0} \frac{a^3}{4\pi r^3} - 0.36} \tag{4.104}$$

If the spheres are small compared to the period and the period is small compared to the wavelength, the reflection coefficient is normally small, since $ka \ll 1$ and $r \ll a$, so the imaginary term in (4.103) is large. However, the array is highly reflective if the expression in brackets in formula (4.103) is small. This phenomenon is well known for single spheres, and it is called the *Fröhlich mode* resonance [28]. Formula (4.103) shows how the particle interaction affects this phenomenon. For large distances between the inclusions $(a \gg r)$, the effect takes place for $\epsilon = -2\epsilon_0$, which is the result for individual spheres [28]. With increasing density of the grid, the resonant value of ϵ changes and the effect becomes more pronounced.

As another example, consider an array of short-wire scatterers loaded by arbitrary bulk impedances Z_{load} in their centers. The polarizability of a single inclusion is (see Chapter 5)

$$\alpha = \left(\frac{4Z_{\text{inp}} + Z_{\text{load}}}{Z_{\text{inp}} + Z_{\text{load}}} \right) \frac{l^2}{3j\omega Z_{\text{inp}}} \tag{4.105}$$

where Z_{inp} is the input impedance of the wire antenna, and l is the half-length of the wire. In this case, parameter $2Z_g/\eta$ in the expression for the reflection coefficient (4.101) reads

$$\frac{2Z_g}{\eta} = \frac{2a^2}{\eta} \left[\frac{3Z_{\text{inp}}}{l^2} \frac{(Z_{\text{inp}} + Z_{\text{load}})}{(4Z_{\text{inp}} + Z_{\text{load}})} - \frac{\eta \epsilon_0 \mu_0}{6\pi} \omega^2 \right] + \frac{j}{2} \left(\frac{\cos kR_0}{kR_0} - \sin kR_0 \right) \tag{4.106}$$

For lossless scatterers, this simplifies to

$$\frac{2Z_g}{\eta} = j \left[\frac{6a^2}{\eta l^2} \operatorname{Im} \left\{ Z_{\text{inp}} \frac{Z_{\text{inp}} + Z_{\text{load}}}{4Z_{\text{inp}} + Z_{\text{load}}} \right\} + \frac{1}{2} \left(\frac{\cos kR_0}{kR_0} - \sin kR_0 \right) \right] \tag{4.107}$$

which is always imaginary. This result can be used in the design and modeling of tunable and adaptive thin sheets.

4.6   DOUBLE ARRAYS OF SCATTERERS

Various arrays are often located near conducting surfaces or near interfaces with dielectric media. Multilayer arrays are also of considerable interest.

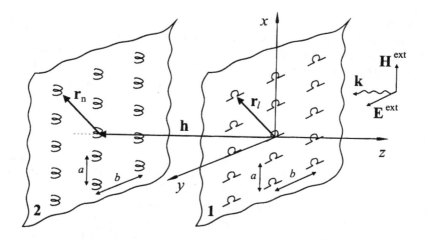

Figure 4.16 Two parallel arrays of small particles. In each particle, both electric and magnetic moments can be induced. We neglect the higher-order multipoles.

We need to model these structures in the design of absorbing coverings, frequency selective surfaces, or photonic (electromagnetic) stopband materials (Chapter 6). The same problem must be solved to understand arrays near metal bodies. The key configuration to study here is the system of two parallel arrays of particles, as illustrated in Figure 4.16. We already know how small electromagnetic scatterers interact within a single planar array. To understand layered configurations, we will next consider the interaction field between two parallel arrays [29, 30].

4.6.1 Interaction Between Parallel Arrays

Considering the geometry shown in Figure 4.16, our main goal is to design a model for the fields created by the particles in one array at the positions of the particles in the other array. The particles are excited by local electric and magnetic fields $\mathbf{E}^{\text{loc}}$ and $\mathbf{H}^{\text{loc}}$ that are the sums of the external fields and the fields created by the array dipoles:

$$\mathbf{E}_1^{\text{loc}} = \mathbf{E}_1^{\text{ext}} + \overline{\overline{\beta}}_e(0) \cdot \mathbf{p}_1 + \overline{\overline{\beta}}_{ee}(h) \cdot \mathbf{p}_2 + \overline{\overline{\beta}}_{em}(h) \cdot \mathbf{m}_2$$

$$\mathbf{E}_2^{\text{loc}} = \mathbf{E}_2^{\text{ext}} + \overline{\overline{\beta}}_e(0) \cdot \mathbf{p}_2 + \overline{\overline{\beta}}_{ee}(h) \cdot \mathbf{p}_1 + \overline{\overline{\beta}}_{em}(h) \cdot \mathbf{m}_1$$

$$\mathbf{H}_1^{\text{loc}} = \mathbf{H}_1^{\text{ext}} + \overline{\overline{\beta}}_m(0) \cdot \mathbf{m}_1 + \overline{\overline{\beta}}_{mm}(h) \cdot \mathbf{m}_2 + \overline{\overline{\beta}}_{me}(h) \cdot \mathbf{p}_2$$

$$\mathbf{H}_2^{\text{loc}} = \mathbf{H}_2^{\text{ext}} + \overline{\overline{\beta}}_m(0) \cdot \mathbf{m}_2 + \overline{\overline{\beta}}_{mm}(h) \cdot \mathbf{m}_1 + \overline{\overline{\beta}}_{me}(h) \cdot \mathbf{p}_1$$

(4.108)

Here, indices $1, 2$ refer to the two arrays. The system is investigated in such frequency ranges that the distance between the elements is smaller than approximately $\lambda/2$, where λ is the wavelength of the external field. On the other hand, the characteristic dimensions of the particles are smaller than the distance between them. In view of these conditions we assume that the scatterers can be replaced by electric or magnetic dipoles, or a combination of them. The distance between the array planes is h, and vector $\mathbf{h}$ is parallel to the axis z. The system is under normal plane-wave excitation. We know how to estimate the interaction fields in single arrays, and our next goal is to estimate the interaction coefficients $\overline{\overline{\beta}}(h)$ modeling field interactions between two parallel arrays.

Interaction Field

This problem can be solved using a generalization of the approach described in Section 4.5. Under the normal plane-wave excitation all the dipoles in each array are equal and in-phase, and the geometry of the problem is such that the local field acting on the particles in one plane is the same for all of them. This allows us to place the origin of the Cartesian coordinate system (x, y, z) at the center of an arbitrary particle and restrict the analysis to the local field for the particle at the origin and for the particle at $(0, 0, -h)$. To find the interaction field avoiding direct summation, we choose a circle of radius R on the planes of both arrays, and for each plane replace the discrete dipole moment distribution outside the circle by a continuous distribution of dipole moments, as in modeling single arrays in Section 4.5. The continuous distributions (surface densities of dipole moments) are connected to the discrete dipole moments as

$$\mathbf{p}_s = \frac{\mathbf{p}}{S_0}, \qquad \mathbf{m}_s = \frac{\mathbf{m}}{S_0} \qquad (4.109)$$

where $\mathbf{p}_s$ and $\mathbf{m}_s$ denote the electric and magnetic dipole surface densities, respectively. S_0 is the cell area ($S_0 = ab$ for rectangular cells, Figure 4.16). Next, from every sum in the expression for the interaction field we extract the contributions of distant dipoles ($r_i > R$). These extracted sums we replace by integrals that are the fields of the homogenized dipole moment distributions. As before, we denote by $\mathbf{E}^{\text{sheet}}$ the field created by the continuous current sheet with a hole of radius R, at the axis of the hole and at a distance h from its center (Figure 4.17). The total interaction field between the planes is the sum of $\mathbf{E}^{\text{sheet}}$ and the field created by the discrete dipoles inside the circle.

To evaluate the integrals for $\mathbf{E}^{\text{sheet}}$ and $\mathbf{H}^{\text{sheet}}$, we split the electric and magnetic dipole moments of the particles into components that are parallel and orthogonal to the array planes and consider them separately. After some algebra, the integrals reduce to simple combinations of elementary functions

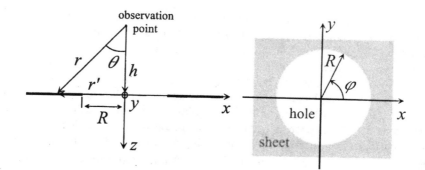

Figure 4.17 Calculation of the interaction field between two parallel arrays. The observation point is at distance h from the array plane. The field of a homogenized current sheet with a hole of radius R is calculated.

[29]. For codirected dipoles parallel to the array plane, we find:

$$\mathbf{E}^{\text{sheet}}(\mathbf{p}, h) = -j\omega \mathbf{p}_s \frac{\eta}{4} \left\{ 1 - \frac{1}{jk\sqrt{R^2 + h^2}} \right.$$

$$\left. + \frac{h^2}{R^2 + h^2} \left(1 + \frac{1}{jk\sqrt{R^2 + h^2}} \right) \right\} e^{-jk\sqrt{R^2+h^2}} \qquad (4.110)$$

$$\mathbf{E}^{\text{sheet}}(\mathbf{m}, h) = \frac{j\omega}{2} \frac{\mathbf{h} \times \mathbf{m}_s}{\sqrt{R^2 + h^2}} e^{-jk\sqrt{R^2+h^2}} \qquad (4.111)$$

For dipoles orthogonal to the plane, we have:

$$\mathbf{E}^{\text{sheet}}(\mathbf{p}, h) = -\frac{j\omega \mathbf{p}_s \eta}{2} \left\{ 1 + \frac{1}{jk\sqrt{R^2 + h^2}} \right\} \frac{R^2}{R^2 + h^2} e^{-jk\sqrt{R^2+h^2}} \qquad (4.112)$$

$$\mathbf{E}^{\text{sheet}}(\mathbf{m}, h) = \mathbf{H}^{\text{sheet}}(\mathbf{p}, h) = 0 \qquad (4.113)$$

Magnetic fields created by magnetic dipoles can be found by replacing $\mathbf{p}_s$ by $\mathbf{m}_s$ and η by $1/\eta$. For $h = 0$, these results reduce to that of Section 4.5. We see from the above expressions that for $h \to \infty$, they give simple plane-wave fields created by the averaged polarizations. Also for $R = 0$ we have a uniformly polarized sheet (electric or magnetic current sheet), and the results correctly give the plane-wave field excited by a current plane.

Approximate Analytical Formulas and the Floquet Series

We know that for dense arrays only a few nearly located inclusions must be considered as separate scatterers in the calculations of the interaction field, and the main contribution can be evaluated as that from the averaged dipole moment density. As in Section 4.5, let us make use of our freedom to

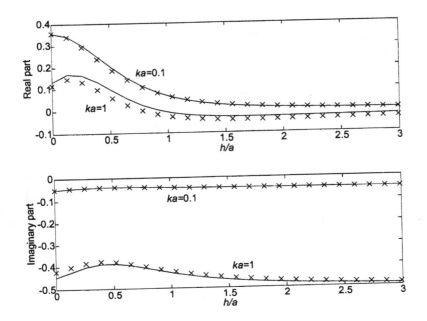

Figure 4.18 Real and imaginary parts of the normalized electric field $E\epsilon_0 a^3 \exp(jkh)/p$ as functions of h/a. Electric dipoles are parallel to the array plane. Exact (solid lines) and approximate ($\times$) results.

choose the radius R, inside which we consider the inclusions individually. If we now replace all the particles of a square cell array except the reference one by a "sheet with a hole" (which means we choose the hole radius R to be smaller than the distance between the particles a), we can find such a radius of the hole that the results given for $h = 0$ by (4.110) and (4.112) coincide at zero frequency with the known static results. For $h = 0$, this takes place for $R = R_0 = a/(4C_{par})$ for dipoles parallel to the plane and $R_0 = a/(2C_{ort})$ for dipoles orthogonal to the plane, where C_{par} and C_{ort} are the known interaction constants at zero frequency [23]. Numerical studies show that substitution of the same effective radius into the formulas for the interaction fields for $h \neq 0$ gives quite good results as well. Curves calculated by this approximate method are compared with the exact solution in Figures 4.18 and 4.19 (surprisingly good agreement with the exact solution is seen). We can observe how for small values of ka, the real part of the interaction constant for the electric field approaches its static value for planar arrays (0.36).

In principle, we could establish an even more accurate model defining the effective hole radius as a function of the distance h to the array plane. However, the calculated results show that there would be only a small im-

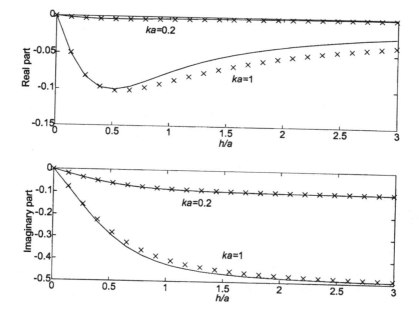

Figure 4.19 The same exact and approximate results as in Figure 4.18 for the normalized magnetic field $H\eta\epsilon_0 a^3 \exp(jkh)/p$.

provement but the model becomes far more complicated. This approximate method gives reasonable results for grids with square cells only. The reason is that in other cases the round shape of the hole does not correspond to the cell shape.

The imaginary part of the interaction constants describing electromagnetic interactions between two arrays can be determined exactly in closed form [31], similar to single arrays of dipoles. To find the imaginary part of $\beta(h)$ that corresponds to the contribution of the second array into the interaction field at the first array, we consider an auxiliary problem. Let an array of lossless electric dipole scatterers be placed in front of a magnetic wall at distance $h/2$ from the wall. Then there is an image array located at distance h from the original array. Its dipoles are cooriented to the actual dipoles. If α' is the particle polarizability *in array* that takes into account the screen influence (the influence of the mirror array), the imaginary part of $1/\alpha'$ can be obtained in the same manner as in Section 4.5.2 for a single particle. The external field power for this case reads:

$$P^{\text{ext}} = -\frac{\omega}{2}\,\text{Im}(\alpha')|E^{\text{ext}}|^2 \qquad (4.114)$$

Here the external field is $E^{\text{ext}} = 2\cos(kh/2)E^{\text{inc}}$, due to the screen influence (we position the phase center on the screen surface). The power radiated back from one cell of the array reads:

$$P^{\text{rad}} = \frac{1}{2} S_0 \left\{ \eta \widehat{J} e^{-jkh/2} \cos(kh/2) \right\} \left\{ \widehat{J} e^{-jkh/2} \cos(kh/2) \right\}^*$$

$$= \frac{\eta \omega^2 |p|^2}{2 S_0} \cos^2(kh/2) = \frac{\eta \omega^2 |\alpha'|^2}{2 S_0} \cos^2(kh/2) |E^{\text{ext}}|^2 \qquad (4.115)$$

Here, S_0 is the cell area. In the lossless case $P^{\text{rad}} = P^{\text{ext}}$, and we find that

$$\text{Im} \left\{ \frac{1}{\alpha'} \right\} = \frac{\eta \omega}{S_0} \cos^2(kh/2) \qquad (4.116)$$

For the considered case the polarizability in array α' is

$$\alpha' = \frac{\alpha}{1 - \alpha \beta(0) - \alpha \beta(h)}, \qquad \text{or} \qquad \frac{1}{\alpha'} = \frac{1}{\alpha} - \beta(0) - \beta(h) \qquad (4.117)$$

Here, $\beta(0)$ is the interaction constant for a single array and $\beta(h)$ is the interaction constant between the two arrays. From (4.87), (4.116), and (4.117) we obtain

$$\text{Im} \left\{ \frac{1}{\alpha'} \right\} = \frac{\eta \omega}{S_0} \cos^2(kh/2) = \frac{\eta \omega}{2 S_0} - \text{Im}\{\beta(h)\} \qquad (4.118)$$

Hence, we can write

$$\text{Im}\{\beta(h)\} = -\frac{\eta \omega}{2 S_0} \cos(kh) \qquad (4.119)$$

This is an exact formula for the imaginary part of the interaction constant $\beta(h)$. The physical meaning of (4.119) is very simple: This value when multiplied by the dipole moment p gives the imaginary part of the plane wave field generated by the array at distance h. Note that the obtained restriction is valid in the general case, not only for an array near a magnetic screen.

Finally, we arrive at approximate analytical expressions for the interaction constant between two arrays with square cells [29]:

$$\overline{\overline{\beta}}_{ee}(h) = - \text{Re} \left[\frac{jw\eta}{4a^2} \left\{ 1 - \frac{1}{jk\sqrt{R_0^2 + h^2}} \right. \right.$$

$$+ \frac{h^2}{R_0^2 + h^2} \left(1 + \frac{1}{jk\sqrt{R_0^2 + h^2}} \right) \left. \right\} e^{-jk\sqrt{R_0^2 + h^2}} \overline{\overline{I}}_t +$$

$$\frac{jw\eta}{2a^2} \left\{ 1 + \frac{1}{jk\sqrt{R_0^2 + h^2}} \right\} \frac{R_0^2}{R_0^2 + h^2} e^{-jk\sqrt{R_0^2 + h^2}} \frac{\mathbf{hh}}{h^2} + \qquad (4.120)$$

$$\frac{1}{4\pi\epsilon_0} \left\{ \frac{1}{h^3} + \frac{jk}{h^2} - \frac{k^2}{h} \right\} e^{-jkh} \overline{\overline{I}}_t - \frac{1}{2\epsilon_0} \left\{ \frac{1}{h^3} + \frac{jk}{h^2} \right\} \frac{\mathbf{hh}}{h^2} \right] - j\frac{\eta\omega}{2a^2} \cos kh \, \overline{\overline{I}}_t$$

$$\overline{\overline{\beta}}_{em}(h) = \frac{j\omega}{2a^2} \frac{\mathbf{h} \times \overline{\overline{I}}_t}{\sqrt{R_0^2 + h^2}} e^{-jk\sqrt{R_0^2+h^2}} \tag{4.121}$$

$$\overline{\overline{\beta}}_{mm}(h) = \overline{\overline{\beta}}_{ee}(h)/\eta^2, \qquad \overline{\overline{\beta}}_{me}(h) = -\overline{\overline{\beta}}_{em}(h) \tag{4.122}$$

Here $\mathbf{h}$ is a vector orthogonal to the source array. It is directed from the observation point (the point where the field is to be determined) to the array plane, and $|\mathbf{h}| = h$ (Figure 4.17). Parameter R_0 is equal to $a/1.438$, and $\overline{\overline{I}}_t = \overline{\overline{I}} - \mathbf{hh}/h^2$. The imaginary parts of the interaction dyadics $\overline{\overline{\beta}}_{ee}$ and $\overline{\overline{\beta}}_{mm}$ are exact. The real parts are approximate, and the model is accurate for $ka < 1 \ldots 1.5$.

As an alternative, the Floquet series can be used to estimate $\overline{\overline{\beta}}(h)$. In contrast to the calculation of the interaction field for a single array, when the Floquet mode expansion cannot be used, it is a very helpful tool for modeling interactions between two arrays. Applying the Poisson summation rule (4.11) to the sum of dipole fields (twice for this double sum), we arrive at the following result (see the theory of infinite antenna arrays, e.g., in [4]):

$$\mathbf{E}(\mathbf{p}, h) = -\frac{j}{2\epsilon_0} \sum_{m=-\infty}^{+\infty} \sum_{l=-\infty}^{+\infty} \left[k^2 - \left(\frac{2\pi m}{a}\right)^2 \right] \frac{e^{-j\sqrt{k^2 - \left(\frac{2\pi m}{a}\right)^2 - \left(\frac{2\pi l}{a}\right)^2} |h|}}{\sqrt{k^2 - \left(\frac{2\pi m}{a}\right)^2 - \left(\frac{2\pi l}{a}\right)^2}} \mathbf{p}_s \tag{4.123}$$

where the square root branch is defined by $\mathrm{Im}\sqrt{\cdot} < 0$. Equation (4.123) is for dipoles parallel to the grid plane and square-cell grids. This series can be used for practical calculations if the distance between the grids h is larger than the grid period a. For large h, only one leading term gives a very accurate result (obviously, this leading term $m = l = 0$ is simply the plane wave generated by the averaged polarization). Expressions like (4.110) are reasonable to use in calculations of the layer field at positions near (as compared to the period a) to the array plane, where the Floquet expansion converges very slowly. On the other hand, the closed-form analytical model is less accurate for large distances from the layer plane. Although that formula also gives the plane-wave field as the leading contribution, the correction to that field at large distances is given more accurately by the Floquet expansion formula (4.123).

4.6.2 Reflection and Transmission Through Double Arrays

The model of the interaction field for the system of two parallel dipole arrays allows calculation of the reflection and transmission coefficients in a simple analytical form. Let us consider a system of two parallel square-cell arrays of magnetically polarizable dipole particles (solution for arrays of electrical dipoles can be found from the duality principle). The geometry of the problem is depicted in Figure 4.16. The system is illuminated by a plane wave $\mathbf{H}^{\text{ext}} = \mathbf{H}_0 e^{jkz}$. The particles of the two arrays have scalar magnetic polarizabilities that are equal to α_1 and α_2, where the index refers to the first and the second array, respectively. Induced magnetic dipole moments of the particles are determined by the values of the local fields:

$$\mathbf{m}_1 = \alpha_1 \mathbf{H}_1^{\text{loc}}, \qquad \mathbf{m}_2 = \alpha_2 \mathbf{H}_2^{\text{loc}} \tag{4.124}$$

Using the approximate formulas for the interaction field (see Sections 4.5 and 4.6.1) we can write the local fields in the following form:

$$\mathbf{H}_1^{\text{loc}} = \mathbf{H}_0 + \beta(0)\mathbf{m}_1 + \beta(h)\mathbf{m}_2$$

$$\mathbf{H}_2^{\text{loc}} = \mathbf{H}_0 e^{-jkh} + \beta(0)\mathbf{m}_2 + \beta(h)\mathbf{m}_1 \tag{4.125}$$

where the interaction coefficients $\beta(0)$ and $\beta(h)$ are given by (4.90) and (4.120)[8]. Solving (4.124) with (4.125), we find

$$\mathbf{m}_1 = \frac{\mathbf{H}_0}{\Delta} \left\{ \frac{1}{\alpha_2} - \beta(0) + \beta(h)e^{-jkh} \right\} \tag{4.126}$$

$$\mathbf{m}_2 = \frac{\mathbf{H}_0}{\Delta} \left\{ \left(\frac{1}{\alpha_1} - \beta(0) \right) e^{-jkh} + \beta(h) \right\} \tag{4.127}$$

where $\Delta = [\alpha_1^{-1} - \beta(0)][\alpha_2^{-1} - \beta(0)] - [\beta(h)]^2$. To calculate the far-zone reflected or transmitted field, the parallel-oriented magnetic dipoles belonging to one plane can be replaced by a magnetic current sheet carrying magnetic surface current $\hat{\mathbf{J}}_m = j\omega\mathbf{m}/S_0$. The magnetic field created by this current sheet is $\mathbf{H} = -\hat{\mathbf{J}}_m e^{-jk|z|}/2\eta$, for the case when the sheet is located at $z = 0$. Now we can write the reflected field as a sum of two parts created by the first and the second arrays:

$$\mathbf{H}^{\text{ref}} = \mathbf{H}_1^{\text{ref}} + \mathbf{H}_2^{\text{ref}} = -\frac{j\omega}{2\eta S_0} \left(\mathbf{m}_1 + \mathbf{m}_2 e^{-jkh} \right) \tag{4.128}$$

Hence, the reflection coefficient is

$$R = \frac{H^{\text{ref}}}{H^{\text{ext}}} \bigg|_{z=0} = -\frac{j\omega}{2\eta S_0 \Delta} \left\{ \frac{e^{-j2kh}}{\alpha_1} + \frac{1}{\alpha_2} - \beta(0)(1 + e^{-j2kh}) + 2\beta(h)e^{-jkh} \right\} \tag{4.129}$$

The transmission coefficient reads

$$T = 1 - \frac{j\omega}{2\eta S_0 \Delta} \left\{ \frac{1}{\alpha_1} + \frac{1}{\alpha_2} - 2\beta(0) + 2\beta(h) \cos(kh) \right\} \tag{4.130}$$

Models for polarizabilities of various particles are needed to make actual calculations. Such models will be introduced in the next chapter.

---

[8]To find the interaction constants for magnetic particles, the results given by (4.90) and (4.120) should be divided by η^2.

4.6.3   Reflection from an Array Near a Metal Screen

The above solution can be used to find the reflection coefficient from an array of magnetic particles near a metal screen. This problem is relevant to the design of radar absorbers based on artificial magnetics. Let us consider the case of one array of magnetic dipole particles with scalar polarizabilities α. The array plane is located at distance $h/2$ from a metal screen. The incident field is the same as assumed in the previous section: $\mathbf{H}^{\text{ext}} = \mathbf{H}_0 e^{jkz}$. In considering this problem, we can use Figure 4.16 assuming that the first plane is the plane of the dipole scatterers and the second plane is the mirror image of those scatterers in the ideally conducting plane. The value of the magnetic dipole moment of the particles is

$$\mathbf{m} = \alpha \mathbf{H}^{\text{loc}} \tag{4.131}$$

where the local field can be written in terms of the incident field $\mathbf{H}_0$ and the interaction field $\mathbf{H}^{\text{int}}$:

$$\mathbf{H}^{\text{loc}} = \mathbf{H}_0(1 + e^{-jkh}) + \mathbf{H}^{\text{int}} \tag{4.132}$$

The factor $1 + e^{-jkh}$ is needed to account for the reflection of the incident wave in the screen. The interaction field $\mathbf{H}^{\text{int}}$ is proportional to the magnetic moment of the particles:

$$\mathbf{H}^{\text{int}} = \beta \mathbf{m} \tag{4.133}$$

Here, the interaction coefficient β takes into account all the interactions of the particles in the system, so it equals $\beta = \beta(0) + \beta(h)$. After simple algebra, we find the induced dipole moment:

$$\mathbf{m} = \frac{\alpha \left(1 + e^{-jkh}\right)}{1 - \beta\alpha} \mathbf{H}_0 \tag{4.134}$$

Next, we find the reflected magnetic field:

$$\mathbf{H}^{\text{ref}} = \mathbf{H}_0 e^{-jkh} - \frac{j\omega \mathbf{m}}{2\eta S_0}\left(1 + e^{-jkh}\right) \tag{4.135}$$

and the reflection coefficient

$$R = \frac{H^{\text{ref}}}{H^{\text{ext}}}\bigg|_{z=0} = e^{-jkh} - \frac{j\omega\alpha\left(1 + e^{-jkh}\right)^2}{2\eta S_0(1 - \beta\alpha)} \tag{4.136}$$

The analytical result for the reflection coefficient allows us to define requirements for the optimum values of particle polarizabilities needed to achieve specific goals. Obviously, the reflective properties strongly depend on the interaction constant that defines parameter β in (4.136). In Chapter 6 we will use these results to model thin artificial impedance layers for applications in antenna techniques and as radar absorbers.

4.7 DIFFRACTION BY EDGES OF WIRE GRIDS

Wire grids and meshes are widely used in antenna techniques. Regular planar infinite wire grid screens can be easily analyzed numerically, since the periodicity of the structure allows us to reduce the computational domain to only one period, applying the periodical Green function technique. However, large finite and curved structures can be studied only with the help of approximate techniques such as the averaged boundary conditions. In this section, we demonstrate how this technique is used in studies of more complicated and realistic systems.

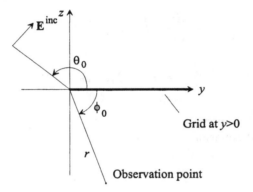

Figure 4.20 Semi-infinite wire grid, located at $z = 0$, $y > 0$ illuminated by a plane wave. Wires of the grid are orthogonal to the grid edge. The incidence angle is θ_0.

For example, in the design of antenna reflectors it is necessary to understand how electromagnetic waves diffract at the edges of grid reflectors. Let us consider a semi-infinite wire grid illuminated by a plane wave [32] (Figure 4.20). The incident field is

$$H_x^{inc} = H_0 e^{jk(z \sin \theta_0 + y \cos \theta_0)}$$

$$E_y^{inc} = H_x^{inc} \eta \sin \theta_0, \qquad E_z^{inc} = -H_x^{inc} \eta \cos \theta_0 \qquad (4.137)$$

If the period of the grid is small compared with the wavelength, the averaged boundary condition (4.39) can be used to model the wire array:

$$E_y^{inc} + E_y = j \frac{\eta}{2} \alpha_{ABC} \left(J_y + \frac{1}{k^2} \frac{d^2 J_y}{dy^2} \right), \qquad z = 0, \quad y > 0 \qquad (4.138)$$

Here, J_y is the averaged current density induced on the grid, and E_y is the y-component of the electric field created by this current. This field we calculate in the standard way, introducing the vector potential:

$$\mathbf{E} = jk\eta\left(\mathbf{A} + \frac{1}{k^2}\nabla\nabla\cdot\mathbf{A}\right), \qquad \mathbf{H} = \nabla\times\mathbf{A} \qquad (4.139)$$

where

$$\mathbf{A} = \frac{j}{4}\int_0^\infty J_y(y')H_0^{(2)}[k\sqrt{z^2+(y-y')^2}]\,dy' \qquad (4.140)$$

Substituting into the boundary condition (4.138), we arrive at

$$\frac{k}{4}\left(1 + \frac{1}{k^2}\frac{d^2}{dy^2}\right)\int_0^\infty J_y(y')H_0^{(2)}(k|y-y'|)\,dy'$$

$$= \begin{cases} -\frac{j}{2}\alpha_{\mathrm{ABC}}\left(1 + \frac{1}{k^2}\frac{d^2}{dy^2}\right)J_y(y) + H_0\sin\theta_0 e^{jky\cos\theta_0}, & y > 0 \\ \mathcal{E}(y), & y < 0 \end{cases} \qquad (4.141)$$

where $\mathcal{E}(y)$ is an unknown function proportional to E_y at $z = 0$, $y < 0$.

This integral equation can be solved using the factorization technique. Applying the Fourier transform

$$f(w) = \int_{-\infty}^\infty f(y)e^{-jwy}\,dy \qquad (4.142)$$

to (4.141), the integral equation is transformed to the functional equation

$$\frac{\alpha_{\mathrm{ABC}}}{2k^2}(k^2-w^2)\Psi(w)J_-(w) = -\frac{H_0\sin\theta_0}{w - k\cos\theta_0} - j\mathcal{E}_+(w) \qquad (4.143)$$

where

$$\Psi(w) = 1 - \frac{jk}{\alpha\sqrt{k^2-w^2}} \qquad (4.144)$$

We have assumed that the wavenumber k has a small negative imaginary part and $\pi/2 < \theta_0 < \pi$ (that is, $\cos\theta_0 < 0$), and used (4.12) for the Fourier transform of the Hankel function. The Fourier transformed current density $J_-(w)$ is an analytical function in the lower half plane of the complex variable w ($\mathrm{Im}\,w < 0$), and $\mathcal{E}_+(w)$ is analytical in the upper half plane. Factorization of function $L(w) = (k^2 - w^2)\Psi(w) = L_+(w)L_-(w)$ into two functions, such that one is analytical in the upper half plane of w and the other one in the lower half plane, is known (e.g., [33]). Next, we divide (4.143) by $L_+(w)$:

$$\frac{\alpha_{\mathrm{ABC}}}{2k^2}L_-(w)J_-(w) = -\frac{H_0\sin\theta_0}{(w - k\cos\theta_0)L_+(w)} - j\frac{\mathcal{E}_+(w)}{L_+(w)} \qquad (4.145)$$

The left-hand side is analytical in the lower half plane, and function $\mathcal{E}_+/L_+$ is analytical in the upper half plane, and we only should express the remaining term, proportional to the incident field amplitude, as a sum of two functions analytical in the lower and upper half planes, respectively. This can be done using the following identity:

$$\frac{1}{(w - k\cos\theta_0)L_+(w)} = \frac{1}{(w - k\cos\theta_0)L_+(k\cos\theta_0)}$$

$$+ \frac{1}{w - k\cos\theta_0}\left[\frac{1}{L_+(w)} - \frac{1}{L_+(k\cos\theta_0)}\right] \quad (4.146)$$

The first term on the right-hand side is analytical in the lower half plane, and the second one in the upper half plane. Now we can write (4.143) in an equivalent form as

$$\frac{\alpha_{ABC}}{2k^2}L_-(w)J_-(w) + \frac{H_0\sin\theta_0}{(w - k\cos\theta_0)L_+(k\cos\theta_0)}$$

$$= -\frac{H_0\sin\theta_0}{w - k\cos\theta_0}\left[\frac{1}{L_+(w)} - \frac{1}{L_+(k\cos\theta_0)}\right] - j\frac{\mathcal{E}_+(w)}{L_+(w)} \quad (4.147)$$

Here, the function on the left-hand side is analytical in the lower half plane and the right-hand side function is analytical in the upper half plane. Thus, each of the two functions equals the same constant, which is zero because the current density vanishes at the grid edge: $J(0) = 0$. This determines the Fourier transform of the induced current:

$$J_-(w) = -\frac{2k^2 H_0\sin\theta_0}{\alpha_{ABC}L_-(w)L_+(k\cos\theta_0)(w - k\cos\theta_0)} \quad (4.148)$$

Finally, the inverse Fourier transform expressing the field scattered and reflected by the grid can be calculated (in the far zone) using the steepest descent method (see [32]). Even more complicated problems can be solved analytically with the help of the averaged boundary conditions, for example diffraction by grid edges if the wires are oriented at an arbitrary angle with respect to the array edge [34].

Another powerful technique for the analysis of diffraction from impedance wedges and half planes is the Malyuzhinets technique [35, 36].

Problems

4.1 Calculate the radar cross section of a planar rectangular reflector made of a dense wire grid (normal incidence, the size of the reflector is much larger than the wavelength).

4.2 Compare the approximate value of the interaction constant (4.75) with the exact value (4.89).

4.3 Check that (4.82) holds for the Hertzian dipole.

4.4 Analyze reflection properties of a regular dense grid of small loaded dipole antennas. Use (4.105) and substitute realistic antenna parameters. Find the resonance conditions for the reflection coefficient.

4.5 Study how the Fröhlich-mode resonance frequency and the resonance curve width for an array of dielectric spheres depend on the array period.

4.6 Consider an array of magnetically polarizable and lossy particles near a metal screen as an antireflective coating. Find the required particle polarizability.

References

[1] Munk, B.A., *Frequency Selective Surfaces: Theory and Design,* New York: John Wiley & Sons, 2000.

[2] Wu, T.K., (ed.), *Frequency Selective Surface and Grid Array,* New York: John Wiley & Sons, 1995.

[3] Felsen, L.B., and N. Marcuvitz, *Radiation and Scattering of Waves,* Piscataway, NJ: IEEE Press, 1991.

[4] Hansen, R.C., *Phased Array Antennas,* New York: Wiley-Interscience, 1997.

[5] Kontorovich, M.I., et al., *Electrodynamics of Grid Structures,* Moscow: Radio i Svyaz, 1987 (in Russian).

[6] Kontorovich, M.I., et al., "Reflection Factor of a Plane Electromagnetic Wave Reflecting from a Plane Wire Grid," *Radio Eng. & Electron Phys.,* No. 2, 1962, pp. 222-231.

[7] Vainshtein, L.A., "On the Electromagnetic Theory of Gratings. I. Ideal Grating in Free Space," *High-Power Electronics, Issue 2,* pp. 26-56, Moscow: Nauka, 1963 (in Russian) [English translation in *High-Power Electronics,* Oxford: Pergamon Press, 1966].

[8] Marcuvitz, N., (ed.), *Waveguide Handbook,* Lexington, MA: Boston Technical Publishers, 1964.

[9] Yatsenko, V.V., et al., "Higher Order Impedance Boundary Conditions for Sparse Wire Grids," *IEEE Trans. Antennas and Propagation,* Vol. 48, No. 5, 2000, pp. 720-727.

[10] Lamb, H., "On the Reflection and Transmission of Electric Waves by a Metallic Grating," *Proc. London Math. Soc.*, Vol. 29, Ser. 1, 1898, pp. 523-544.

[11] MacFarlane, G.G., "Surface Impedance of an Infinite Parallel-Wire Grid at Oblique Angles of Incidence," *J. IEE*, Vol. 93 (III A), No. 10, 1946, pp. 1523-1527.

[12] DeLyser, R.R., and E.F. Kuester, "Homogenization Analysis of Electromagnetic Strip Gratings," *J. Electromagnetic Waves and Applications*, Vol. 5, No. 11, 1991, pp. 1217-1236.

[13] DeLyser, R.R., "Use of the Boundary Conditions for the Solution of a Class of Strip Grating Structures," *IEEE Trans. Antennas and Propagation*, Vol. 41, No. 1, 1993, pp. 103-105.

[14] Whites, K.W., and R. Mittra, "An Equivalent Boundary-Condition Model for Lossy Planar Periodic Structures at Low Frequencies," *IEEE Trans. Antennas and Propagation*, Vol. 44, No. 12, 1996, pp. 1617-1629.

[15] Wait, J.R., "Reflection at Arbitrary Incidence from a Parallel Wire Grid," *Appl. Sci. Res.*, Vol. B IV, 1954, pp. 393-400.

[16] Wait, J.R., "Reflection from a Wire Grid Parallel to a Conducting Plane," *Can. J. Phys.*, Vol. 32, 1954, pp. 571-579.

[17] Wait, J.R., "The Impedance of a Wire Grid Parallel to a Dielectric Interface," *IRE Trans. Microwave Theory and Techniques*, Vol. MTT-5, No. 2, 1957, pp. 99-102.

[18] Young, J.L., and J.R. Wait, "Note on the Impedance of a Wire Grid Parallel to a Homogeneous Interface," *IEEE Trans. Microwave Theory and Techniques*, Vol. 37, No. 7, 1957, pp. 1136-1138.

[19] Adonina, A.I., and V.V. Shcherbak, "Equivalent Boundary Conditions at a Metal Grating Situated Between Two Magnetic Materials," *Zh. Tekh. Fiz.*, Vol. 34, 1964, pp. 333-335 (in Russian; English translation in *Sov. Phys. Tech. Phys.*, Vol. 9, 1964, pp. 261-263).

[20] Gradshteyn, I.S., and I.M. Ryzhik, *Table of Integrals, Series, and Products*, 6th ed., San Diego, CA: Academic Press, 2000.

[21] Serdyukov, A., et al., *Electromagnetics of Bi-Anisotropic Materials: Theory and Applications*, Amsterdam: Gordon and Breach Science Publishers, 2001.

[22] Maslovski, S.I., and S.A. Tretyakov, "Full-Wave Interaction Field in Two-Dimensional Arrays of Dipole Scatterers," *Int. J. Electronics and Communications (AEÜ)*, Vol. 53, No. 3, 1999, pp. 135-139.

[23] Collin, R.E., *Field Theory of Guided Waves,* 2nd ed., Piscataway, NJ: IEEE Press, and Oxford, England: Oxford University Press, 1991.

[24] Jackson, J.D., *Classical Electrodynamics,* 3rd ed., New York: John Wiley & Sons, 1999.

[25] Grimes, C.A., and D.M. Grimes, "The Effective Permeability of Granular Thin Films," *IEEE Trans. Magnetics,* Vol. 29, 1993, pp. 4092-4094.

[26] Tretyakov, S.A., et al., "Impedance Boundary Conditions for Regular Dense Arrays of Dipole Scatterers," *Electromagnetics Laboratory Report Series,* Helsinki University of Technology, Report 304, August 1999, to appear in *IEEE Trans. Antennas and Propagation.*

[27] Landau, L.D., and E.M. Lifshitz, *Statistical Physics,* 3rd ed., Part 1, Oxford, England: Butterworth-Heinemann, 1997, pp. 377-379.

[28] Bohren, G.F., and D.R. Huffman, *Absorption and Scattering of Light by Small Particles,* New York: John Wiley & Sons, 1983, pp. 139-140.

[29] Yatsenko, V.V., S.I. Maslovski, and S.A. Tretyakov, "Electromagnetic Interaction of Parallel Arrays of Dipole Scatterers," *Progress in Electromagnetics Research,* Vol. 25, pp. 285-307, 2000.

[30] Yatsenko, V.V., et al., "Plane-Wave Reflection from Double Arrays of Small Magnetoelectric Scatterers," *IEEE Trans. Antennas and Propagation,* Vol. 51, No. 1, pp. 2-11, 2003.

[31] Yatsenko, V.V., and S.I. Maslovski, "Electromagnetic Diffraction by Double Arrays of Dipole Scatterers," *Proceedings of the International Seminar Day on Diffraction,* St. Petersburg, 1999, pp. 196-209.

[32] Rozov, V.A., and S.A. Tretyakov, "Diffraction of Plane Electromagnetic Waves by a Semi-Infinite Grid of Parallel Conductors," *Radioengineering and Electronic Physics,* Vol. 26, No. 11, 1981, pp. 6-15.

[33] Weinstein, L.A., *The Theory of Diffraction and the Factorization Method,* Boulder, CA: Golem Press, 1969 (translation from Russian by P. Beckmann).

[34] Rozov, V.A., and S.A. Tretyakov, "Diffraction of Plane Electromagnetic Waves by a Semi-Infinite Grid Made of Parallel Conductors Arranged at an Angle to the Grid's Edge," *Radioengineering and Electronic Physics,* Vol. 29, No. 5, 1984, pp. 37-47.

[35] Osipov, A.V., and A.N. Norris, "The Malyuzhinets Theory for Scattering from Wedge Boundaries: A Review," *Wave Motion,* Vol. 29, 1999, pp. 313-340.

[36] Norris, A.N., and A.V. Osipov, "Far-Field Analysis of the Malyuzhinets Solution for Plane and Surface Waves Diffraction by an Impedance Wedge," *Wave Motion*, Vol. 30, 1999, pp. 69-89.

Chapter 5

Composite Materials

Most generally defined, materials are "the matter from which a thing is or can be made" [1]. Speaking about materials for applications in radio and microwave engineering, we usually distinguish between dielectrics, magnetics, metals, semiconductors, and superconductors. What are *composite materials* and *metamaterials*? It appears that there is no clear and more or less universally adopted definition, especially for the relatively new term *metamaterial*. A *composite* is something "made up of recognizable constituents" [1], but for our purposes this definition is too general and loose, because we can say so about just any material (composed of atoms). In material science, *composite* is "a multiphase material formed from a combination of materials which differ in composition or form, remain bonded together, and retain their identities and properties" [2]. In electromagnetics literature this is equivalent to *inhomogeneous* materials, sometimes also called *composite media*. Let us adopt this understanding of the term: If we combine two or more materials (say, dielectric or metal inclusions embedded in another dielectric) to produce another material (effectively behaving as another dielectric or an artificial magnetic, for example), this new material is a composite. Here we will consider both random distribution of dielectric or metal particles and regular periodic lattices.

5.1 CONSTITUENTS OF COMPOSITES AND METAMATERIALS: THIN-WIRE INCLUSIONS

Radio engineers have been using artificial dielectrics for a long time. In these materials, the role of molecules in usual dielectrics is played by small dielectric or metal spheres, dispersed in a host medium (usually a low-loss and low-permittivity dielectric). At radio frequencies, the molecules of conventional dielectrics are very small compared with the wavelength, which usually means that they weakly respond to electromagnetic fields. That is why it is an advantage to use larger scatterers like metal spheres.

If we want to have more freedom in the design and control of the inclusion properties, the spherical shape is clearly not the best choice. Making small particles of thin conducting wires, it is possible to vary the inclusion shape, and moreover, it is possible to include nonlinear devices and small electronic circuits into each composite inclusion. Another advantage is a possibility to reduce the resonance frequency of a particle without increasing the volume occupied by the particle. This makes the design more compact. Composite materials for microwave and radio frequency applications contain inclusions that are large enough to be manufactured from ordinary copper wires, or using planar technology, as three-dimensional printed circuit boards. The basic building block of such metamaterials is a piece of thin conducting wire of a simple shape. From the point of view of the design of metamaterials, it is of prime interest to consider two basic shapes: a straight wire and a loop. This is because a short straight conducting wire polarizes in external electric fields as an electric dipole, and a small loop creates a magnetic dipole. Mixtures of these basic particles allow us to create artificial magnetodielectrics, and their properties can be engineered, if we learn how to control the electromagnetic response of these two artificial "molecules." Connecting a short straight wire dipole and a small wire loop together in a single inclusion allows us to design more complicated artificial media: reciprocal bianisotropic composites. Here, the two basic shapes are the canonical helix [3] and the omega particle [4]. We will present a simple analytical model for wire scatterers, possibly loaded by some bulk impedances in the center. The loads not only allow us to control the inclusion properties, but also provide models of more complex particles such as chiral and omega inclusions.

5.1.1 Polarization of Loaded Wire Dipoles

Let us consider a metal wire with the total length $2l$ and wire radius $r_0 \ll l$ loaded by a bulk impedance Z_{load} in its center (see Figure 5.1). We are mostly interested in the analytical modeling of the polarization of the wire at low frequencies when the particle is much smaller than the wavelength ($2l \ll \lambda$) by an external electromagnetic field E_{inc} directed along the wire axis. We need to know the electric dipole moment induced in this "artificial molecule" by the external field. An estimation can be found in [5], where the role of the load is played by a conducting loop antenna. However, the model of short loaded wires developed in [5] assumes the linear current distribution along the wire, which is not always a good approximation. The model explained here is free from this assumption. Although in many applications the inclusions are small compared to the wavelength, we will first develop a model without this restriction and then consider the simplified formulas for small particles.

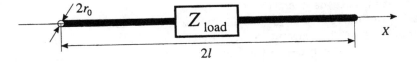

Figure 5.1 Geometry of a loaded wire.

As is well known in antenna theory (e.g., [6]), in the transmitting regime the current distribution along a thin wire excited by a point voltage source at its center is approximately sinusoidal:

$$I_{\text{trans}}(x) = I_{\text{trans}}(0) f_t(x), \qquad f_t(x) = \frac{\sin k(l - |x|)}{\sin kl} \qquad (5.1)$$

and in the receiving mode, the current distribution for a short-circuited antenna is given by

$$I_{\text{rec}}(x) = I_{\text{rec}}(0) f_r(x), \qquad f_r(x) = \frac{\cos kx - \cos kl}{1 - \cos kl} \qquad (5.2)$$

Here, $I_{\text{trans}}(0)$ and $I_{\text{rec}}(0)$ are the current amplitudes at the antenna center that depend on the source power, and k is the wavenumber in the surrounding medium. The distribution of current along a wire antenna illuminated by an electromagnetic wave and loaded by an arbitrary load is a combination of these two functions. In the low-frequency limit the current distributions are modeled by simpler functions:

$$f_t(x) \approx 1 - \frac{|x|}{l}, \qquad f_r(x) \approx 1 - \frac{x^2}{l^2} \qquad (5.3)$$

that are approximations of (5.1) and (5.2).

For a receiving antenna loaded by Z_{load} in its center, we can find the current in the antenna center as

$$I(0) = \frac{\mathcal{E}}{Z_{\text{load}} + Z_{\text{inp}}} = \frac{1}{Z_{\text{load}} + Z_{\text{inp}}} \int_{-l}^{l} E_{\text{inc}} f_t(x) \, dx \qquad (5.4)$$

Here, Z_{inp} is the input impedance of the wire antenna and $\mathcal{E}$ is the electromotive force induced by the incident field.

Because of the key role of (5.4), let us now give its derivation. Equation (5.4) follows from the Lorentz lemma

$$\int (\mathbf{E}_2 \cdot \mathbf{J}_1 - \mathbf{E}_1 \cdot \mathbf{J}_2) \, dV \qquad (5.5)$$

that is valid for the electric fields $\mathbf{E}_1$ and $\mathbf{E}_2$ created, respectively, by two sources with current densities $\mathbf{J}_1$ and $\mathbf{J}_2$ in a reciprocal medium. Here the integration extends over the whole space. Consider an arbitrary antenna loaded by impedance Z_{load} and illuminated by electric field $\mathbf{E}_{\text{inc}}$. This external field and the current induced by this field in the load will be the first set of fields and sources in the Lorentz lemma ($\mathbf{E}_1$ and $\mathbf{J}_1$). Consider also the same antenna in the transmitting mode (excited by a voltage source V_0 with the internal impedance equal to the load impedance Z_{load}), and denote the density of the current generated by that source on the antenna wire as $\mathbf{J}$. That will be the second set of vectors in the Lorentz lemma. The input current in case of the transmitting antenna is

$$I_{\text{inp}} = \frac{V_0}{Z_{\text{load}} + Z_{\text{inp}}} \tag{5.6}$$

where Z_{inp} is the input impedance of the transmitting antenna. Applying the Lorentz lemma and using (5.6), we can write

$$\int \mathbf{E}_{\text{inc}} \cdot \mathbf{J}\, dV = V_0 I = I_{\text{inp}}(Z_{\text{load}} + Z_{\text{inp}})I \tag{5.7}$$

From here we find the current in the load of the *receiving* antenna I:

$$I = \frac{1}{Z_{\text{load}} + Z_{\text{inp}}} \int \mathbf{E}_{\text{inc}} \cdot \mathbf{f}_t\, dV \tag{5.8}$$

where $\mathbf{f}_t = \mathbf{J}/I_{\text{inp}}$ is the normalized current density distribution function in the transmitting regime. Obviously, the integral gives us the electromotive force induced by the incident field on the antenna load. We stress that it is determined by the distribution of the current on the antenna *in the transmitting mode* (the distribution function $\mathbf{f}_t = \mathbf{J}/I_{\text{inp}}$). If the external electric field is constant along the wire, the integral expressing the electromotive force in (5.4) reads

$$\mathcal{E} = E_{\text{inc}} \frac{2\tan(kl/2)}{k} \approx E_{\text{inc}}l \tag{5.9}$$

where the approximation is valid for $|k|l \ll 1$.

Let us rewrite (5.4) as

$$I(0) = \frac{\mathcal{E}}{Z_{\text{inp}}} - \frac{\mathcal{E}}{(Z_{\text{load}} + Z_{\text{inp}})} \frac{Z_{\text{load}}}{Z_{\text{inp}}} = \frac{\mathcal{E}}{Z_{\text{inp}}} - \frac{Z_{\text{load}}I(0)}{Z_{\text{inp}}} \tag{5.10}$$

In this form, the total current is presented as a superposition of two contributions: the first term is the current induced in the short-circuited antenna by the incident field (5.4) with $Z_{\text{load}} = 0$, and the second term is the current generated by an equivalent voltage source $V_0 = -Z_{\text{load}}I(0)$ at the position of the load. The first part of the current is induced by the incident field directly, and the second part is "scattered" by the antenna load.

Now we can apply (5.1) and (5.2) and find the total current distribution along the loaded wire. Indeed, the first component in the load current is generated by the incident field directly (amplitude proportional to $\mathcal{E}$); thus, its distribution along the wire is that of the antenna in the receiving regime. The other component is scattered by the load, and its current distribution is that of the transmitting antenna. This determines the current at an arbitrary position x:

$$I(x) = \frac{\mathcal{E}}{Z_{\text{inp}}} f_r(x) - \frac{\mathcal{E}}{(Z_{\text{load}} + Z_{\text{inp}})} \frac{Z_{\text{load}}}{Z_{\text{inp}}} f_t(x) \qquad (5.11)$$

Consideration of two limiting cases is instructive. If $Z_{\text{load}} \to 0$, we get

$$I(x) = I(0) f_r(x) \qquad (5.12)$$

as it should be in the short-circuit case. If $Z_{\text{load}} \to \infty$, we have

$$I(x) = \frac{\mathcal{E}}{Z_{\text{inp}}} [f_r(x) - f_t(x)] \qquad (5.13)$$

Note that in this case $I(0) = 0$. Equation (5.13) can be better understood in the case of short antennas. Assuming the simple approximations (5.3) for the current distributions, we get for $Z_{\text{load}} \to \infty$

$$I(x) = \frac{E}{Z_{\text{inp}}} \frac{|x|}{l} \left(1 - \frac{|x|}{l}\right) = \frac{E}{4Z_{\text{inp}}} \left[1 - \left(1 - \frac{|x|}{l/2}\right)^2\right] \qquad (5.14)$$

This is an expected result: The current is zero at the center and of course at the wire ends. Along the two arms there is a quadratic distribution with the quarter amplitude at the two separate parts of the antenna. The charge is distributed according to a linear law, and the distribution is the same on the two arms. The current distributions are illustrated by Figure 5.2.

The knowledge of the current distribution makes it possible to calculate the induced electric dipole moment p:

$$p = \int_{-l}^{l} q(x) x \, dx = -\frac{1}{j\omega} \int_{-l}^{l} \frac{dI(x)}{dx} x \, dx = \frac{1}{j\omega} \int_{-l}^{l} I(x) \, dx \qquad (5.15)$$

where $q(x)$ is the charge distribution. After a simple integration we find the polarizability α defined as $\alpha = p/E_{\text{inc}}$:

$$\alpha = \left[\frac{\sin(kl)/k - l\cos kl}{1 - \cos kl} - \frac{1 - \cos kl}{k\sin kl} \frac{Z_{\text{load}}}{Z_{\text{inp}} + Z_{\text{load}}}\right] \frac{4\tan(kl/2)}{j\omega k Z_{\text{inp}}} \qquad (5.16)$$

Here we have assumed that E_{inc} is uniform over the wire volume and used (5.9). In the quasistatic regime, when $|k|l \ll 1$, the formula for the polarizability (5.16) simplifies to

$$\alpha = \left(\frac{2}{3} - \frac{1}{2} \frac{Z_{\text{load}}}{Z_{\text{inp}} + Z_{\text{load}}}\right) \frac{2l^2}{j\omega Z_{\text{inp}}} = \left(\frac{4Z_{\text{inp}} + Z_{\text{load}}}{Z_{\text{inp}} + Z_{\text{load}}}\right) \frac{l^2}{3j\omega Z_{\text{inp}}} \qquad (5.17)$$

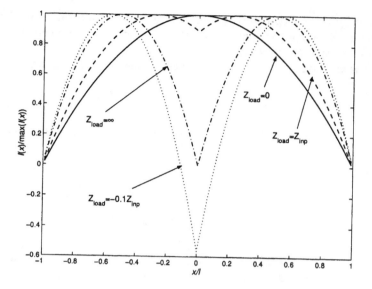

Figure 5.2 Distribution of current along a short wire scatterer for different load impedances Z_{load}.

For $Z_{\text{load}} = -Z_{\text{inp}}$, as expected, we have a resonance where in the absence of any losses the polarizability becomes infinitely high. Close to the resonance, the current distribution along short wires is nearly triangular, since the second term in (5.11) dominates. This is the reason why a simple model of [5] that assumes the triangular current distribution gives very good estimations near the particle resonance.[1] In addition, we can observe that for $Z_{\text{load}} = -4Z_{\text{inp}}$ the polarizability becomes zero (this takes place due to a specific current distribution; see Figure 5.3). Thus, it is possible to effectively tune the polarizability of a wire by varying its load. For small ideally conducting particles loaded by purely reactive loads the polarizability is a real function (no absorption), and we see that it changes sign depending on the load and the frequency.

Next, we will discuss the particles of the other basic shape: circular loaded loops. Such inclusions are used to provide magnetic response of composites and metamaterials.

5.1.2 Polarization of Loaded Loops

Magnetic response of composites and metamaterials can be realized using small loaded loop antennas. Consider a small loop formed by an ideally

---

[1] The resonant region is the most interesting for applications because the particles can be strongly excited.

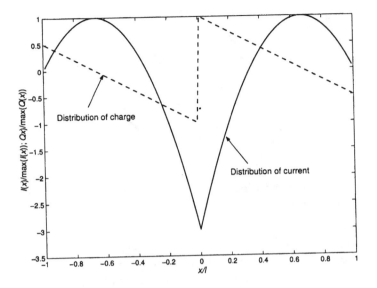

Figure 5.3 Distributions of charge and current for wire scatterers loaded by $Z_{\text{load}} = -4Z_{\text{inp}}$. In this case the total dipole moment of the particle is zero.

conducting wire. The current induced in the loop positioned in an external magnetic field H orthogonal to the loop plane is the ratio of the electromotive force induced in the loop and the loop impedance:

$$I = \frac{-j\omega\mu_0 S}{Z_{\text{loop}}} H = -\frac{\mu_0 S}{L} H \qquad (5.18)$$

Here, S is the loop area and $Z_{\text{loop}} = j\omega L$ is the inductive loop impedance. The magnetic moment of the loop is $m = \mu_0 IS = -\mu_0^2 S^2 H/L$. Notice the negative sign in this formula: Due to the Lenz law, the induced current creates magnetic flux that *decreases* the incident flux. From the point of view of material design, this means that closed loop particles can be used in the design of artificial diamagnetics with $0 < \mu_{\text{eff}} < \mu_0$. To create a paramagnetic inclusion, we can load the loop by a capacitance. In this case, the current in the loop becomes

$$I = \frac{-j\omega\mu_0 S}{j\omega L + \frac{1}{j\omega C}} H = \frac{\omega^2 \mu_0 SC}{1 - \omega^2 LC} H \qquad (5.19)$$

The coefficient in this equation changes sign at the resonance frequency $\omega_0 = \sqrt{1/(LC)}$: This particle is paramagnetic at frequencies below the resonance and diamagnetic above the resonance.

Let us now analyze small loaded loop scatterers in more detail. The geometry of the particle is shown in Figure 5.4. Let us assume that the

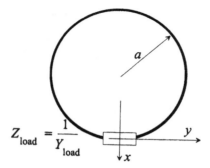

Figure 5.4 Geometry of a loaded loop. The loop is formed by a conducting wire of radius $r_0 \ll a$.

electrical size of the loop is not large, more precisely, that $|k|a < 1$, where $k = \omega\sqrt{\epsilon\mu}$ is the wavenumber in the background medium. Our goal here is to find the magnetic and electric dipole moments induced in this particle by external electromagnetic fields. This can be done [5] using the theory of loop antennas. For moderate-sized particles[2] we can use analytical expressions for the antenna parameters.

Circular loop antennas are analyzed with the use of the Fourier series expansion of the current distribution function with respect to the polar angle ϕ measured around the loop. The exact solution for the Fourier coefficients is given, for example, in [7, Chapter 9]. The analytical model of a circular loop antenna of moderate size can be constructed by keeping only the most important terms in the expansion, which results in the following approximate formula for the input admittance [8]

$$Y_{\text{loop}} = \frac{-j}{\pi\eta}\left(\frac{1}{A_0} + \frac{2}{A_1}\right) \tag{5.20}$$

with the first two Fourier coefficients

$$A_0 = \frac{ka}{\pi}\left[\log(8a/r_0) - 2\right] + \frac{1}{\pi}\left[0.667(ka)^3 - 0.267(ka)^5\right]$$

$$-j\left[0.167(ka)^4 - 0.033(ka)^6\right]$$

$$A_1 = \left(ka - \frac{1}{ka}\right)\frac{1}{\pi}\left[\log(8a/r_0) - 2\right] + \frac{1}{\pi}\left[-0.667(ka)^3 + 0.207(ka)^5\right]$$

$$-j\left[0.333(ka)^2 - 0.133(ka)^4 + 0.026(ka)^6\right] \tag{5.21}$$

---

[2]When the size becomes large compared to the wavelength, the very concept of effective media governed by local averaged material equations becomes invalid.

This expression corresponds to the approximation of the loop current distribution in the form

$$I(\phi) = \frac{-jV_0}{\pi\eta}\left(\frac{1}{A_0} + \frac{2\cos\phi}{A_1}\right) \tag{5.22}$$

where V_0 is the electromotive force of an ideal voltage point source positioned at the polar angle $\phi = 0$ (the position of the bulk load in Figure 5.4). The loop antenna input impedance is the inverse of its admittance: $Z_{\text{loop}} = 1/Y_{\text{loop}}$. The expression for the input antenna admittance (5.20) is valid for antennas in lossy media, as well as in free space.

Polarization by Magnetic Fields

Suppose that the particle is set in a high-frequency magnetic field orthogonal to the loop plane. The uniform part of the receiving loop current is given by [9]:

$$I_0 = -2a\frac{J_1(ka)}{A_0}\left(1 + \frac{j}{Y_{\text{loop}} + Y_{\text{load}}}\frac{1}{\pi\eta A_0}\right)H_z \tag{5.23}$$

where H_z is the z-component of the external magnetic field and J_1 is the Bessel function of the first order. The corresponding magnetic dipole moment m is proportional to I_0: $m = \mu\pi a^2 I_0$, and the polarizability is

$$a_{mm} = \frac{m}{H_z} = -2\mu\pi a^3\frac{J_1(ka)}{A_0}\left(1 + \frac{j}{Y_{\text{loop}} + Y_{\text{load}}}\frac{1}{\pi\eta A_0}\right) \tag{5.24}$$

Polarizability of a closed loop that is a classical result [10], follows from (5.24) as a special case of infinite input admittance of the bulk load Y_{load}.

Because the loop is not uniform but there is a load at one point, the current distribution around the loop is essentially nonuniform even for small loops. The main nonuniform component is the "$\cos\phi$"-component of the current. This nonuniformity of the current means accumulation of charge at the opposite sides of the loop; thus, an electric dipole directed along the y-axis is induced by external magnetic field H_z. The corresponding current amplitude is

$$I_1 = -4ja\frac{J_1(ka)}{A_0}\frac{1}{\pi\eta A_1}\frac{1}{Y_{\text{loop}} + Y_{\text{load}}}H_z \tag{5.25}$$

that gives the cross-polarizability component $a_{em}^{(yz)} = p_y/H_z$:

$$a_{em}^{(yz)} = -\frac{4a^2}{\omega\eta}\frac{J_1(ka)}{A_0 A_1}\frac{1}{Y_{\text{loop}} + Y_{\text{load}}} \tag{5.26}$$

In this notation, the lower index *em* means that there is an electric response to magnetic fields, and the upper index means that while the exciting field vector is directed along z, the induced dipole points along y.

We see that magnetic fields induce both magnetic and electric dipole moments in loaded loops. Such particles are called *bianisotropic particles* [11]. They can be characterized by dyadic electric and magnetic polarizabilities that define the bianisotropic relations between induced electric and magnetic dipole moments $\mathbf{p}$, $\mathbf{m}$ and local electric and magnetic fields $\mathbf{E}$, $\mathbf{H}$:

$$\mathbf{p} = \bar{\bar{a}}_{ee} \cdot \mathbf{E} + \bar{\bar{a}}_{em} \cdot \mathbf{H} \tag{5.27}$$

$$\mathbf{m} = \bar{\bar{a}}_{me} \cdot \mathbf{E} + \bar{\bar{a}}_{mm} \cdot \mathbf{H} \tag{5.28}$$

Due to the reciprocity principle, the polarizability dyadics are subject to the relations [11]:

$$\bar{\bar{a}}_{em} = -\bar{\bar{a}}_{me}^T, \qquad \bar{\bar{a}}_{ee} = \bar{\bar{a}}_{ee}^T, \qquad \bar{\bar{a}}_{mm} = \bar{\bar{a}}_{mm}^T \tag{5.29}$$

where T denotes the transpose dyadic. For the simple loop geometry, the polarizability dyadics can be assumed to have the following forms

$$\bar{\bar{a}}_{ee} = a_{ee}^{(xx)} \mathbf{x}_0 \mathbf{x}_0 + a_{ee}^{(yy)} \mathbf{y}_0 \mathbf{y}_0 \tag{5.30}$$

$$\bar{\bar{a}}_{mm} = a_{mm} \mathbf{z}_0 \mathbf{z}_0 \tag{5.31}$$

$$\bar{\bar{a}}_{em} = a_{em}^{(yz)} \mathbf{y}_0 \mathbf{z}_0 \tag{5.32}$$

$$\bar{\bar{a}}_{me} = a_{me}^{(zy)} \mathbf{z}_0 \mathbf{z}_0 \tag{5.33}$$

$(a_{em}^{(yz)} = -a_{me}^{(zy)})$. Here we assume that the wire is thin, so that its radius r_0 is much smaller than the loop radius a, and the polarization in electric fields oriented orthogonally to the loop plane can be neglected.

Excitation by Electric Fields Polarized in the Loop Plane

If the incident electric field is in the plane of the loop, two electric field polarizations have to be distinguished, since the loop response depends on the orientation of the field with respect to the position of the load. The current excited in the loop can be divided into two parts, so that one of them does not depend on the load position and coincides with that excited in a closed loop. For the case when the external electric field is polarized along the x-axis (Figure 5.4), this component actually represents the whole current. It can be expressed as [8, 9]

$$I_1(\phi) = j \frac{4a J_1'(ka)}{\eta A_1} \sin \phi \, E_x \tag{5.34}$$

$[J_1'(ka)$ is the Bessel function derivative]. Here, we assume the external field to be polarized along the x-axis without any loss of generality. Equation (5.34) allows us to find the polarizability integrating the corresponding charge distribution:

$$a_{ee}^{(xx)} = -\frac{4\pi a^2 J_1'(ka)}{\omega \eta A_1} \tag{5.35}$$

The second component of the current can be considered as one generated by an equivalent source located at the load position, and that component is proportional to the y-component of the external electric field. For the electric field polarized along the axis y, the total current in the loop can be written as

$$I(\phi) = \frac{4aJ_1'(ka)}{A_1} \left[\frac{-j}{\eta} \cos\phi + \frac{\epsilon}{\pi\mu} \frac{1}{(Y_{\text{loop}} + Y_{\text{load}})} \left(\frac{1}{A_0} + \frac{2\cos\phi}{A_1} \right) \right] E_y \tag{5.36}$$

The uniform part of the current (5.36) generates a magnetic dipole along the z-axis, and the cross-polarizability $a_{me}^{(zy)}$ can then be easily found as

$$a_{me}^{(zy)} = \frac{4\epsilon a^3 J_1'(ka)}{A_0 A_1} \frac{1}{Y_{\text{loop}} + Y_{\text{load}}} \tag{5.37}$$

The terms in (5.36) that vary as $\cos\phi$ correspond to the polarizability component $a_{ee}^{(yy)}$. Due to the discontinuity at the load, that polarizability contains an additional term, as compared to $a_{ee}^{(xx)}$:

$$a_{ee}^{(yy)} = -\frac{4\pi a^2 J_1'(ka)}{\omega\eta A_1} \left(1 + \frac{2j}{\pi\eta A_1} \frac{1}{Y_{\text{loop}} + Y_{\text{load}}} \right) \tag{5.38}$$

Electrically Small Loaded Loop

For electrically small loops the input admittance (5.20) (and the corresponding impedance) can be modeled by simple lumped-element circuits. The input admittance of the loop antenna can be approximated as [7,9]

$$Y_{\text{loop}} = \frac{j\omega C_1}{1 + j\omega C_1 R_d} + \frac{1}{j\omega L_0 + R_l} \tag{5.39}$$

The equivalent parameters in (5.39) read [6,7,9]:

$$L_0 = \mu a \left(\log \frac{8a}{r_0} - 2 \right) \tag{5.40}$$

$$C_1 = \frac{\epsilon \pi^2 a}{3 \log(a/r_0)} \tag{5.41}$$

$$R_l = (Sk^2)^2 \frac{\eta}{6\pi} \tag{5.42}$$

$$R_d = (ak)^2 \frac{\eta}{6\pi} \tag{5.43}$$

where $S = \pi a^2$ is the loop area.

These relations can be obtained directly from (5.20) as limiting cases at low frequencies. For the loop antenna, the uniform component of the current distribution defines the loop inductance L_0, and the dipole-mode current component corresponds to the additional loop capacitance C_1 and the radiation resistance R_d [7,9]. Alternatively, the additional loop capacitance can be found by calculation of the electric field energy stored in the loop field [6]. The two approaches give close but still different results: As follows from (5.20), the additional capacitance $C_1 = 2a\epsilon/[\log(8a/r_0) - 2]$, and the result of [6] is given by (5.41).

The polarizabilities of small loops can be calculated with the use of the simplified relations for particles that are small compared to the wavelength. The loop current distribution coefficients $1/A_0$ and $1/A_1$ that appear in the expressions for some of the polarizabilities can be approximated as

$$\frac{1}{A_0} \approx \frac{\pi}{ka\,[\log(8a/r_0) - 2]}\left\{1 + j\frac{\pi}{6}\frac{(ka)^3}{[\log(8a/r_0) - 2]}\right\} \qquad (5.44)$$

$$\frac{1}{A_1} \approx -\frac{\pi ka}{\log(8a/r_0) - 2}\left\{1 - j\frac{\pi}{3}\frac{(ka)^3}{[\log(8a/r_0) - 2]}\right\} \qquad (5.45)$$

In this approximation the magnetic polarizability (5.24) reduces to

$$a_{mm}^{(zz)} = -\mu^2\frac{j\omega S^2}{Z_{\text{loop}} + Z_{\text{load}}} \qquad (5.46)$$

This result allows a simple and clear interpretation, as we explained in the beginning of this section. For small loops, the uniform part of the loop current can be found as the electromotive force $\mathcal{E}_B = -j\omega\mu H_z S$ divided by the total impedance $Z_{\text{loop}} + Z_{\text{load}}$. Evaluation of the loop magnetic moment as $m = \mu SI$ leads to (5.46). A similar result was obtained in [6] for a loop antenna loaded by a capacitance.

The polarizability coefficients $a_{em}^{(yz)}$ and $a_{me}^{(zy)}$ simplify to

$$a_{em}^{(yz)} = -a_{me}^{(zy)} = \frac{2\pi^2 a^3\epsilon}{[\log(8a/r_0) - 2]^2}\frac{1}{(Y_{\text{loop}} + Y_{\text{load}})} \qquad (5.47)$$

Electric polarizability $a_{ee}^{(xx)}$ of the small loop can be written as

$$a_{ee}^{(xx)} = \frac{2\pi^2 a^3\epsilon}{\log(8a/r_0) - 2} \qquad (5.48)$$

and assumed to be frequency-independent. Here, the third-order and higher-order terms in the expansion (5.45) may be neglected. A more accurate approach using (5.35) with (5.21) is only needed at frequencies close to the resonance of the loop when its diameter is close to half-wavelength. For the electric field along the y-axis, we arrive at

$$a_{ee}^{(yy)} = \frac{2\pi^2 a^3\epsilon}{\log(8a/r_0) - 2}\left[1 - 2(ka)^2\frac{Z_{\text{load}}}{Z_{\text{loop}} + Z_{\text{load}}}\right] \qquad (5.49)$$

In contrast to the previous polarization, the additional second-order term must be retained since the denominator $Z_{\text{loop}} + Z_{\text{load}}$ becomes very small at the first resonance of the particle.

Connecting a loop and a straight-wire dipole together, we can obtain a simple chiral particle, the so-called *canonical helix* [3]. Its analytical model was developed in [5]. The antenna model of the other canonical shape, the omega particle, can be found in [12].

5.1.3 Excitation of Conducting Particles by External Magnetic Fields

In studies of the polarization of small (characteristic size a) scatterers in electromagnetic fields, it is usually assumed that the exciting fields are uniform in space, although varying in time. This is incompatible with the Maxwell equations: Actually it is implicitly assumed that the fields are uniform only inside the volume occupied by the particle. This approach leads to simple and physically sound results when the particle or its part has the shape of a closed loop (as in Section 5.1.2), when the magnetic field excitation can be easily treated by the calculation of the magnetic flux through the loop area. For arbitrary-shaped particles we should adopt a more realistic model of the fields [13], taking into account that time-varying electromagnetic fields must exhibit variations in space as well.

Calculating the currents excited by nearly uniform magnetic fields, the quasistatic approximation is usually adopted (see [14]). The first task is to find the electric fields that drive the current. Assuming the magnetic induction to be uniform in space, we make use of one of the Maxwell equations in order to evaluate the electric field that drives the currents in conductors:

$$\nabla \times \mathbf{E} = -j\omega \mathbf{B} \tag{5.50}$$

If the magnetic induction is uniform in space, the curl is a constant. If there is a closed loop formed by conductive wire, it is easy to evaluate the induced electromotive force using the Stokes theorem. For arbitrary-shaped inclusions, that is not a trivial task, since the solution of (5.50) in general contains constants that cannot be found without additional physical assumptions. Here we will explain how to treat excitation of arbitrary-shaped (sufficiently small) wire particles in electromagnetic fields.

To model the processes of both the electric-field and magnetic-field excitations, we assume that the particle is positioned in the fields of a standing wave formed by two plane electromagnetic waves of equal amplitudes traveling in the opposite directions of a certain axis z. To see how the particle reacts on electric fields, we position the particle in the maximum of the total electric field of the waves. The electric field of the standing wave depends on the space position according to the following relation:

$$\mathbf{E}(z) = \mathbf{e}_0 E \cos(kz) \qquad (5.51)$$

where $\mathbf{e}_0$ is the unit vector in the direction of the electric field, and E is its complex amplitude measured at the electric field maximum whose position is at $z = 0$. $k = \omega\sqrt{\epsilon_0\mu_0}$ is the wavenumber. If the particle size is small compared to the wavelength, and we keep only the first-order terms $O(ka)$ in our model (remember that a is the characteristic size of the scatterer), we can neglect the space variation, and assume $\mathbf{E}(z) = \mathbf{e}_0 E = \text{const}(z)$. This is possible simply because the Taylor expansion of $\cos(kz)$ has no linear term.

Next, we position the particle in a maximum of the magnetic field of the standing wave in order to see how the particle is excited in magnetic fields. The currents in conducting particles are always excited by electric fields; thus, we should evaluate the electric field of the wave. The electric field in the vicinity of a maximum of the total magnetic field of the standing wave is given by

$$\mathbf{E}(z) = j\eta H \mathbf{z}_0 \times \mathbf{h}_0 \sin(kz) \qquad (5.52)$$

where $\eta = \sqrt{\mu_0/\epsilon_0}$, and $\mathbf{h}_0$ is the unit vector in the direction of the magnetic field. The currents induced in the wire by the field (5.52) can be calculated using conventional methods. Within the frame of the first-order approximation, we replace $\sin(kz)$ by its first term of the Taylor expansion. Obviously, the effect of the magnetic field excitation is a spatial dispersion effect of the first order in ka, where a is the particle size.

Note that in the general case of arbitrary-shaped particles, the currents induced by magnetic fields depend on the direction along which the field varies in space. That means that the very description of the interaction in terms of the magnetic polarizability loses its physical sense. One should instead consider the particle reaction on the spatial derivatives of the electric field. However, note that introducing dependencies on *both* the magnetic field and the spatial derivatives of the electric field in the inclusion model can possibly lead to mistakes.

5.2   MAXWELL GARNETT MODEL

Suppose we know electromagnetic properties of individual inclusions ("artificial molecules"). What will be the properties of a composite material built from these particles? This is a very nontrivial problem, because the answer implies knowledge of how all the particles interact. However, simple and powerful models exist if certain restrictions are satisfied [15, 16]. Let us assume that the particles are small compared to the wavelength and to the average distance between inclusions. In other words, we consider the case of low concentrations of inclusions. In addition, in this section we neglect spatial dispersion effects, assuming that the distance between inclusions is small compared to the wavelength in the medium.

5.2.1 Local and Averaged Fields

The key problem is, of course, the calculation of the local field that excites individual inclusions.[3] As soon as we calculate the local field, we can find the inclusion response multiplying the local field by the known particle polarizability. The local field $\mathbf{E}_{loc}$ exciting one of the particles is the sum of the external field $\mathbf{E}_{ext}$ and the interaction field created by all the other particles in the composite, except the field of the considered particle.[4] Numbering the reference particle by i, we write

$$\mathbf{E}_{loc} = \mathbf{E}_{ext} + \sum_{j \neq i} \mathbf{E}_{ij} \qquad (5.53)$$

Here, $\mathbf{E}_{ij}$ denotes the electric field created by the inclusion number j at the position of the inclusion number i. Obviously, in realistic materials with very many particles, actual calculations using (5.53) are not feasible. Moreover, for the case of an infinite and lossless system, the series in (5.53) has no limit. More precisely, the convergence is not absolute, so the result depends on the order in which the terms are summed up [17]. For limited-size systems, the result depends on the sample shape and size. These problems can be at least partly avoided if we calculate the difference of the local field and the averaged field, instead of the local field as such. This is in harmony with the classical approach to modeling materials that is fundamentally based on a certain averaging of the fields. Averaging allows skipping irrelevant degrees of freedom that describe individual small particles (by considering slowly varying averaged quantities instead of quickly varying in space fields of individual molecules or small particles). This method leads to the formulation of the macroscopic field equations and the effective material relations.

Let us consider the material as a collection of cells, each containing only one particle. The cells are arranged so that the total volume of all cells makes up the total volume of the composite sample. The next step is to perform a spatial average over these cells, which we define as

$$\widehat{\mathbf{E}} = \frac{1}{V} \int_V \mathbf{E}(\mathbf{r})\, dV \qquad (5.54)$$

The integration is performed over the volume of one cell V, and $\mathbf{E}(\mathbf{r})$ is the microscopic electric field.[5] The macroscopic Maxwell equations and the

---

[3]In this calculation we will use some approaches developed by P.A. Belov, C.R. Simovski, and A.P. Vinogradov.

[4]The self-action effect is included in the polarizability that gives the response to the *external* field.

[5]This is not the only possible way to define averaged fields. Various averaging procedures were used in the literature (e.g., [18–20]).

corresponding boundary conditions are formulated for these cell-averaged field quantities. Obviously, the local electric field acting on every inclusion (5.53) is not equal to the averaged electric field (simply because the averaged field includes the field of all particles, but from the local field the reference particle field is excluded).

Consider the difference between the local and averaged fields:

$$\mathbf{E}_{\text{loc}} - \widehat{\mathbf{E}} = \mathbf{E}_{\text{ext}} - \widehat{\mathbf{E}}_{\text{ext}} + \sum_{j \neq i}(\mathbf{E}_{ij} - \widehat{\mathbf{E}}_{ij}) - \widehat{\mathbf{E}}_{ii} \qquad (5.55)$$

First, we observe that in the vast majority of applications all the sources are located far from any point inside the medium as compared to the cell size.[6] If this is the case, then the difference between the external field at any point in the cell and the average of the same field over the cell can be clearly neglected (remember our assumption that the cell size is much smaller than the wavelength). Thus,

$$\mathbf{E}_{\text{ext}} - \widehat{\mathbf{E}}_{\text{ext}} \approx 0 \qquad (5.56)$$

Let us next consider the summation in (5.55). If the particle j is at a large distance from the reference particle i, the difference between the field $\mathbf{E}_{ij}$ at the position of the reference particle i and the same field averaged over its cell $\widehat{\mathbf{E}}_{ij}$ is very small. If all the sizes are small compared to the wavelength, we can use the quasistatic approximation and approximate the fields of distant particles by static fields of electric dipoles. These fields decay as $1/R^3$. The differences between the fields at the cell centers and the cell-averaged fields decay faster, as $1/R^5$. For anisotropic periodical lattices, the summation quickly converges to a constant that depends on the cell geometry plus a small term of the order of $(a/R_{\text{max}})^2$, where a is the cell size and R_{max} is the overall size of the composite sample. If the arrangement of particles is symmetric (centrally positioned particle in regular arrays of small particles arranged in cubical cells), this sum in the quasistatic case tends to zero due to the symmetry of the system. Thus, we conclude that in the quasistatic situation only a few nearest particles influence the difference between the local and the averaged field for any inclusion in the composite. From the mathematical point of view this means that the series in (5.55) absolutely converges, in contrast to the series in (5.53). In case of low-density randomly distributed inclusions we can completely neglect the sum in (5.55), assuming that

$$\sum_{j \neq i}(\mathbf{E}_{ij} - \widehat{\mathbf{E}}_{ij}) \approx 0 \qquad (5.57)$$

This means that it can be neglected for samples that contain many inclusions (assuming that the observation point inside the lattice is located at least 5 to 10 cells away from the sample boundary).

---

[6]Except when there are sources inside the medium or just at its surface. We will discuss these situations later in this chapter.

Naturally, the question arises if this approximation remains valid for larger samples, which can be comparable to or larger than the wavelength. In this situation, wave fields of distant dipoles should be included in the calculation of the sum in (5.55). The wave fields behave as e^{-jkR}/R. Consider for simplicity cubic cells and a wave coming from a distant dipole in the direction orthogonal to one of the cell faces. The cell-averaged field is then

$$\widehat{E}_{ij} \sim \frac{1}{a} \int\limits_{-a/2}^{a/2} \frac{e^{-jkR-jkx}}{R+x}\, dx \approx \frac{e^{-jkR}}{R} \frac{2\sin(ka/2)}{ka} \tag{5.58}$$

The difference

$$E_{ij} - \widehat{E}_{ij} \sim \frac{e^{-jkR}}{R}\left(1 - \frac{2\sin(ka/2)}{ka}\right) \approx \frac{e^{-jkR}}{R}\frac{(ka)^2}{24} \tag{5.59}$$

This quantity is proportional to a very small value of $(ka)^2$ (the cell size is still assumed to be small in wavelengths). However, calculating the sum (5.55) of such terms in the absence of losses in the background medium (real values of k), we arrive at a result which depends on the sample shape and size.[7] As will be obvious from the following, this does not allow introducing effective permittivity of the material because of nonlocality of the electromagnetic response. Losses in the background save the situation, since the influence of distant particles in lossy matrices decays exponentially, so we can consider even infinite media. More discussion on problems related to the homogenization of infinite media can be found in [21]. The volume averaging beyond the quasistatic approximation as a method leading to constitutive relations is also discussed in [22] and references therein.

From these considerations, we conclude that within the frame of the quasistatic approximation the main contribution to the difference between the local and the averaged fields is given by the averaged field of the particle in the reference cell, $\widehat{E}_{ii}$. This is an extremely important result, showing that we deal with a *local* effect. In fact, this is a justification of the very use of the effective material parameters: The effective parameters are independent from the sample size and shape.

Let us calculate the averaged field of one particle over its own cell. Since we are interested in the fields within a unit cell; that is, at small distances from the source, we can assume that the fields obey the quasistatic equations. Following [23, Section 5.5], we consider the particle as a collection of point charges q_m located at positions $\mathbf{r}_m$ inside a spherical cell volume V. The particle field at a point $\mathbf{r}$ within the cell[8] is

$$\mathbf{E}_{ii}(\mathbf{r}) = -\nabla \sum_m \frac{q_m}{4\pi\epsilon_0|\mathbf{r} - \mathbf{r}_m|} \tag{5.60}$$

---

[7] Infinite series of such terms diverges. We have already dealt with similar summations in studies of planar arrays in Chapter 4.

[8] The origin of the coordinate system is at the cell center.

Because our next step is integrating this field over the cell volume, it is convenient to rewrite (5.60) as

$$\mathbf{E}_{ii}(\mathbf{r}) = \sum_m \nabla_m \frac{q_m}{4\pi\epsilon_0 |\mathbf{r} - \mathbf{r}_m|} \tag{5.61}$$

where ∇_m operates on $\mathbf{r}_m$. The averaged field can be now written as

$$\widehat{\mathbf{E}}_{ii} = \frac{1}{V} \sum_m \nabla_m \int_V \frac{q_m}{4\pi\epsilon_0 |\mathbf{r} - \mathbf{r}_m|}\, dV = -\sum_m \mathbf{E}_m \tag{5.62}$$

where V is the cell volume, and

$$\mathbf{E}_m = -\nabla_m \int_V \frac{q_m/V}{4\pi\epsilon_0 |\mathbf{r} - \mathbf{r}_m|}\, dV \tag{5.63}$$

is the field generated at point $\mathbf{r}_m$ by a homogeneous charge distribution with the charge density q_m/V. As such, vectors $\mathbf{E}_m$ satisfy the Poisson equation

$$\nabla_m \cdot \mathbf{E}_m = \frac{1}{\epsilon_0} \frac{q_m}{V} \tag{5.64}$$

This equation can be easily solved. Indeed, if the cell shape is a sphere, the solution can depend only on $\mathbf{r}_m$, and because the right-hand side of (5.64) is a constant, the solution must be a linear function. Thus,

$$\mathbf{E}_m = \frac{1}{3\epsilon_0} \frac{q_m}{V} \mathbf{r}_m \tag{5.65}$$

(we have used the identity $\nabla_m \cdot \mathbf{r}_m = 3$). Finally,

$$\widehat{\mathbf{E}}_{ii} = -\frac{1}{3\epsilon_0 V} \sum_m q_m \mathbf{r}_m = -\frac{1}{3\epsilon_0 V}\mathbf{p} = -\frac{\mathbf{P}}{3\epsilon_0} \tag{5.66}$$

Here, $\mathbf{p} = \sum_m q_m \mathbf{r}_m$ is the electric dipole moment of the inclusion, and $\mathbf{P} = \mathbf{p}/V$ is the average polarization.

The above result has been derived for a spherical cell shape that is a natural choice for random mixtures of inclusions. For regular lattices with cubical cells, the cubical shape of the averaging volume is appropriate. Repeating the same steps for a cubical volume, we find that formulas from (5.60) to (5.64) are valid also in this case (as they are independent from the shape of the volume). However, the solution of (5.64) is not given anymore by (5.65), but for the general shape of the cell we can only write that

$$\mathbf{E}_m = \frac{1}{3\epsilon_0} \frac{q_m}{V} f(\mathbf{r}_m) \tag{5.67}$$

where function $f(\mathbf{r}_m)$ depends on the geometry of the cell. This means that the averaged field $\widehat{\mathbf{E}}_{ii}$ in general is not proportional to the dipole moment of the cell, as in the case of a spherical volume shape.

Equation (5.66), however, remains true if the inclusion is a small dipole particle in the center of a cubical cell. Let us calculate the cell-averaged quasistatic electric field of an electric dipole $\mathbf{p}$ positioned at the center of a cubical cell (side a) (see [24]). The static field of a dipole can be expressed via its scalar potential ϕ:

$$\mathbf{E}_{ii} = -\nabla\phi = -\frac{\mathbf{p}}{4\pi\epsilon_0}\nabla\frac{1}{r^3} \tag{5.68}$$

Directing the axes of a Cartesian coordinate system along the cube faces and positioning the origin at the cube center, for the z-component of $\widehat{\mathbf{E}}_{ii}$ we have:

$$\widehat{E}_z = -\frac{1}{a^3}\int\limits_{-a/2}^{a/2}\int\limits_{-a/2}^{a/2}\int\limits_{-a/2}^{a/2}\frac{\partial\phi}{\partial z}\,dxdydz \tag{5.69}$$

The integration over z is trivial. Because the particle is located at the cell center, the potentials at $z = \pm a/2$ have the same absolute values and the opposite signs, so we have

$$\widehat{E}_z = \frac{-2\widehat{\phi}}{a} \tag{5.70}$$

where $\widehat{\phi}$ denotes the potential of the dipole averaged over the upper wall of the cubic averaging volume:

$$\widehat{\phi} = \frac{1}{a^2}\int\limits_{-a/2}^{a/2}\int\limits_{-a/2}^{a/2}\phi|_{z=a/2}\,dxdy \tag{5.71}$$

Substituting the electric dipole potential, we find

$$\widehat{E}_z = -\frac{1}{2\pi\epsilon_0 a^3}\int\limits_{-a/2}^{a/2}\int\limits_{-a/2}^{a/2}\frac{1}{R^3}\left(p_x x + p_y y + p_z\frac{a}{2}\right)dxdy \tag{5.72}$$

where $R = \sqrt{x^2 + y^2 + (a/2)^2}$. Obviously, x- and y-components of the dipole moment give zero contributions, and after a simple integration

$$I = \int\limits_0^{a/2}\int\limits_0^{a/2}\frac{dxdy}{[x^2 + y^2 + (a/2)^2]^{3/2}} = \frac{\pi}{3a} \tag{5.73}$$

we arrive at

$$\widehat{E}_z = -\frac{p_z}{3\epsilon_0 a^3} \tag{5.74}$$

or, in vector form:

$$\widehat{\mathbf{E}}_{ii} = -\frac{\mathbf{p}}{3\epsilon_0 a^3} = -\frac{\mathbf{P}}{3\epsilon_0} \tag{5.75}$$

which is the same as (5.66). Thus, for cubical cells, this result is valid only if the particle can be replaced by an electric dipole in the cell center. For complex-shaped inclusions this is possible if the inclusion density is low, so that the distance between particles is large compared to the particle size.

Combining (5.55), (5.56), (5.57), and (5.66), we arrive at the famous formula connecting the local and the averaged field, sometimes called the *Lorenz-Lorentz formula*:

$$\mathbf{E}_{\text{loc}} = \widehat{\mathbf{E}} + \frac{\mathbf{P}}{3\epsilon_0} \tag{5.76}$$

The averaged field $\widehat{\mathbf{E}}$, usually denoted simply as $\mathbf{E}$, is the measurable macroscopic field in the medium, and the last formula, giving the field exciting individual inclusions, opens a way to estimate the medium polarization.

5.2.2   Clausius-Mossotti and Maxwell Garnett Formulas

The averaged polarization is

$$\mathbf{P} = \mathbf{p}/V = N\mathbf{p} = N\alpha\mathbf{E}_{\text{loc}} \tag{5.77}$$

where N is the concentration of inclusions (number of inclusions per unit volume), and α is the particle polarizability. Substituting $\mathbf{E}_{\text{loc}}$ from the Lorenz-Lorentz relation (5.76), we get

$$\mathbf{P} = \frac{N\alpha}{1 - \frac{N\alpha}{3\epsilon_0}}\widehat{\mathbf{E}} \tag{5.78}$$

Now we can find the effective permittivity, since by definition

$$\mathbf{D} = \epsilon_0\widehat{\mathbf{E}} + \mathbf{P} = \epsilon_{\text{eff}}\widehat{\mathbf{E}} \tag{5.79}$$

we have

$$\epsilon_{\text{eff}} = \epsilon_0 + \frac{N\alpha}{1 - \frac{N\alpha}{3\epsilon_0}} \tag{5.80}$$

Solving for the polarizability leads to formula

$$\frac{\epsilon_{\text{eff}} - \epsilon_0}{\epsilon_{\text{eff}} + 2\epsilon_0} = \frac{N\alpha}{3\epsilon_0} \tag{5.81}$$

which is called the *Clausius-Mossotti* formula. If the inclusions are dielectric spheres, we can substitute the static polarizability of a sphere

$$\alpha = V(\epsilon - \epsilon_0)\frac{3\epsilon}{\epsilon + 2\epsilon_0} \tag{5.82}$$

(where V is the sphere volume and ϵ is its permittivity) in (5.81), which results in

$$\frac{\epsilon_{\text{eff}} - \epsilon_0}{\epsilon_{\text{eff}} + 2\epsilon_0} = f\frac{\epsilon - \epsilon_0}{\epsilon + 2\epsilon_0} \tag{5.83}$$

where $f = NV$ is the volume fraction of inclusions. This formula bears the name of *Rayleigh* mixing formula. From here we can solve for the effective permittivity:

$$\epsilon_{\text{eff}} = \epsilon_0 + 3f\epsilon\frac{\epsilon - \epsilon_0}{\epsilon + 2\epsilon_0 - f(\epsilon - \epsilon_0)} \tag{5.84}$$

This is the so-called *Maxwell Garnett* formula.

Scattering Loss

Consider equation (5.80) for the effective permittivity:

$$\epsilon_{\text{eff}} = \epsilon_0 + \frac{N\alpha}{1 - \dfrac{N\alpha}{3\epsilon_0}} = \epsilon_0 + \frac{N}{\dfrac{1}{\alpha} - \dfrac{N}{3\epsilon_0}} \tag{5.85}$$

Here, the polarizability α of an individual particle in the composite is a complex number, even if absorption of power in the particle material can be neglected. This is because small particles scatter the power of an incident wave in all directions. This phenomenon is called *scattering loss*, or, when the particle excitation is concerned, *radiation damping*. We know from Chapter 4, (4.82), that to account for the radiation loss we should add the following imaginary number to the inverse value of the particle polarizability:

$$j\,\text{Im}\left\{\frac{1}{\alpha}\right\} = j\frac{k^3}{6\pi\epsilon_0} \tag{5.86}$$

Substituting in (5.85), we have for a composite of lossless (e.g., made of an ideal conductor or a lossless dielectric) inclusions

$$\epsilon_{\text{eff}} = \epsilon_0 + \frac{N}{\text{Re}\left\{\frac{1}{\alpha}\right\} + j\frac{k^3}{6\pi\epsilon_0} - \dfrac{N}{3\epsilon_0}} \tag{5.87}$$

This clearly shows that the composite material is lossy: Its effective permittivity is a complex number.

Consider now a regular lattice of particles modeled by the effective permittivity (5.85). We know that the effective permittivity can be introduced only if the distance between inclusions is considerably smaller than the wavelength, and (5.85) is valid only in this situation. Because the lattice period is considerably smaller than the wavelength, there are no diffraction lobes, meaning that *there is no scattering loss.* Obviously, (5.87) that includes scattering losses is invalid. In our studies of planar arrays, we saw that in regular lattices the scattering loss term was exactly compensated by a part of the interaction field. In the present three-dimensional case the result is the same. We can say that the energy conservation law makes it possible to find the imaginary part of the summation of *dynamic fields* in (5.55): The correct formula for lattices with the period $a < \lambda/2$ is

$$\mathbf{E}_{\text{loc}} - \widehat{\mathbf{E}} = \sum_{j \neq i}(\mathbf{E}_{ij} - \widehat{\mathbf{E}}_{ij}) - \widehat{\mathbf{E}}_{ii} = C\frac{\mathbf{P}}{\epsilon_0} + j\frac{k^3 V}{6\pi\epsilon_0}\mathbf{P} \approx \frac{\mathbf{P}}{3\epsilon_0} + j\frac{k^3 V}{6\pi\epsilon_0}\mathbf{P} \quad (5.88)$$

where C is a real number. Here, k is the wavenumber in the background (matrix) material, not in the effective medium. Direct estimations of this imaginary part (without a recourse to the energy conservation law) show [19, 25] that this is indeed the case. We conclude that for regular lattices of dipole particles (if the lattice period is smaller than $\lambda/2$ in the matrix), the correct estimation for the effective permittivity reads

$$\epsilon_{\text{eff}} = \epsilon_0 + \frac{N}{\frac{1}{\alpha} - j\frac{k^3}{6\pi\epsilon_0} - \frac{N}{3\epsilon_0}} \quad (5.89)$$

If there is no absorption in the particles, the imaginary part of the inverse polarizability is given by (5.86), and the imaginary contributions to the effective permittivity in (5.89) cancel out. If a particle absorbs power, the imaginary part of $1/\alpha$ is larger than (5.86), and the effective permittivity is a complex number accounting for the absorption in the particles.

Next, consider random distributions of small dipole particles in isotropic matrices. In the average, each individual particle feels the interaction field as the field created by the averaged polarization in the medium. But also in regular lattices the contributions from distant inclusions form the same field as the corresponding averaged polarization. The scattering from individual particles in regular lattices is compensated by the interaction field, and we can expect that the situation should be similar in totally random mixtures. If the position of inclusions is random, and on the scale of the wavelength the medium appears as homogeneously polarized, the scattering is also compensated, and (5.89) should be used. Scattering loss in random mixtures is due to fluctuations of the concentration of inclusions. Of course, in any realistic material the particle distribution cannot be neither ideally regular nor totally random, and scattering losses are present, although they can be very small.

Compensation of the radiation damping by the field interaction in optically homogeneous media was probably studied for the first time by L.I.

Mandelshtam. In his polemic with Planck about scattering extinction in the atmosphere [26], he pointed out that if the average distance between molecules is much smaller than the wavelength, the medium should be considered as homogeneous, and the scattering is compensated. It is interesting that in the atmosphere the situation is actually more complicated, because there exist two characteristic length scales: the average distance between molecules and the average free path length. A historical overview of this problem can be found in [27].

Scattering of individual particles is not compensated by the interaction field if the matrix material becomes rather lossy. Indeed, the compensation is due to the electromagnetic interaction between inclusions, and if the waves scattered by the particles substantially decay at the (average) distance between the particles, there will be no compensation of the scattering term. This is typical for absorber applications. In this situation, (5.80) should be used, where the polarizability α includes the scattering loss term. The same is true for samples whose size is smaller than the wavelength [28]. Although even a small piece of a composite material can contain very many particles, in the situation when the overall sample size is smaller than λ, scattering is not compensated, because this compensation is a wave phenomenon. The whole piece of a composite in this case scatters as a single dipole.

5.2.3   Surface Effects

Let us return to the Lorenz-Lorentz formula (5.76) that describes the difference between the local and the averaged fields in bulk structures. From the above discussion, we see that under the specified approximations it is also approximately valid for thin layers, even for cells belonging to the sample boundary [24]. This is a very counterintuitive result because a particle located just on the interface between a medium and free space receives certainly a different influence from its neighbors than a particle in the bulk. Indeed there is a difference, but in regular arrays it is rather small because of the very local nature of the effect. The derivation of (5.76) in Section 5.2.1 shows that the main contribution to the difference between the averaged and local fields in a cell comes from the averaged field of the particle located in that very cell. Despite the smallness of the effect, it can lead to important phenomena (such as the so-called *surface polaritons*), especially in resonant situations. This problem has a long research history [29, 30].

Lorentz Factors for Boundary Layers

To study the surface effects we first turn to the properties of single layers of dipoles. Let us consider an array of parallel electric dipoles forming a square grating at a plane $z = $ const such that the dipole moments **p** are in the array plane.[9] The averaged field in this case is, of course, equal to the external one since the static field of a tangentially polarized dipolar sheet is zero on the sheet plane. If we subtract the external field from the local field, we obtain the interaction field (the contribution of the polarization into the local field), see [31]:

$$\mathbf{E}_{int} = \mathbf{E}_{loc} - \mathbf{E}_{ext} = \frac{C\mathbf{p}}{\epsilon_0 a^3} \qquad (5.90)$$

where a is the grid period. For square arrays of dipoles tangential to the array plane, we have [31, Chapter 12, Formula (27)]:

$$C \approx 0.359 \qquad (5.91)$$

This value of C is different from the estimate $1/3$ obtained for the particles in the bulk. Because $\mathbf{E}_{ext} = \widehat{\mathbf{E}}$, we get

$$\mathbf{E}_{loc} - \widehat{\mathbf{E}} = \frac{0.359\,\mathbf{p}}{\epsilon_0 a^3} \qquad (5.92)$$

For a grating of dipoles orthogonal to the array plane, the result is (5.90) with

$$C \approx -0.719 \qquad (5.93)$$

see formula (30) in Chapter 12 of [31]. (Note that this value actually equals $-2 \cdot 0.359$.) To find the Lorenz-Lorentz relation for this case, we apply the Kontorovich theorem [32], which states that the averaged field of discrete sources equals the field of the averaged sources if the averaging rule is the same (linear but possibly weighted averaging is assumed). Using that theorem, it is easy to show that the field of the grating averaged over a cubic cell $a \times a \times a$ on the grating plane equals the field inside the layer with continuous volume polarization $\mathbf{P} = \mathbf{p}/a^3$:

$$\widehat{\mathbf{E}} = \mathbf{E}_{ext} - \frac{\mathbf{P}}{\epsilon_0} \qquad (5.94)$$

For the difference of the local and averaged fields, we have

$$\mathbf{E}_{loc} - \widehat{\mathbf{E}} = \frac{\mathbf{P}}{\epsilon_0} + \frac{C\mathbf{P}}{\epsilon_0} = \frac{0.281}{\epsilon_0}\mathbf{P} \qquad (5.95)$$

($\mathbf{E}_{loc} = \mathbf{E}_{ext} + \mathbf{E}_{int}$). We observe that the Lorentzian factors are different in these two cases and both are different from the bulk value $1/3$.

Let us now consider an interface between a composite medium and free space. What is the Lorentz factor for the boundary layer? To this end, we study an infinite three-dimensional array of particles. Then the Lorentz

---

[9]Here, we consider the quasistatic approximation. A dynamic model for planar arrays of dipoles is presented in Section 4.5.

factor equals 1/3 for every particle. Let us imagine a plane passing through one of the dipoles so that the dipole lies on the plane. Then we can split the total interaction field created by all the other dipoles into contributions from the dipoles belonging to the same plane (that gives 0.359), and contributions from two half-spaces of dipoles. (Similar reasoning can be found in [30].) Altogether, the resulting factor must be 1/3, thus one half-space brings −0.013. Using this, we can calculate the Lorentz factor for the dipoles on the surface of a half-space. For this case of dipole moments parallel to the interface we have $0.359 − 0.013 = 0.346$.

Consider now the case of dipoles orthogonal to the interface. In this case, deep in the bulk we have the interaction constant $−2/3$, which again gives $1/3 = 1 − 2/3$ for the Lorentz factor. Repeating the above reasoning we get the contribution from a half-space $0.5(−2/3 + 0.719) = 0.0262$. For the boundary layer we have then the Lorentz factor $1 − 0.719 + 0.0262 = 0.3072$.

To calculate the local factor for the second layer below the surface, we need to know the field generated by one plane layer of dipoles at distance a (the array period) from the layer plane. The static solution is available in the literature for dipoles parallel to the array plane (e.g., [30, 31]). The full-wave solution for arbitrary directed dipoles was given above in Section 4.6.1. Applying that theory in the static limit, we find that an array of dipoles parallel to the plane adds −0.01303 into the Lorentz factor, and for the case of perpendicular dipoles the result is 0.02605. Thus, already for the second layers of dipoles we get the Lorentz factors nearly equal to that of the bulk material, which is 1/3. The results are summarized in Table 5.1.

Table 5.1 Lorentz Factors C for Inclusions Close to an Interface

	Parallel polarization	Perpendicular polarization
First layer	0.346	0.307
Second layer	0.3334	0.3332
Bulk material	1/3	1/3

Surface effects are very different at boundaries of regular lattices and random mixtures. The last case was considered in [33], where a model was developed based on the idea that a particle on the boundary has two times less surrounding particles. The reason for the difference is that in the case of regular arrays, the main contribution to the local field comes from the dipoles in the same plane, so it is the same on the boundary as in the bulk. The static field of a regular array of dipoles decreases very fast when going from the array plane, so the rest of the structure gives a small contribution. Naturally, if the surface is rough, then the particles located outside the boundary level experience quite small influence of the bulk, so the averaged Lorentz factor gets much smaller, depending on the distribution factor of the particle position.

Averaged Parameters for Thin Slabs

Consider a thin planar layer of a composite material, which contains only a few layers of inclusions across its thickness. The inclusions are arranged in a regular lattice with square cells. Because the Lorentzian contributions to the local field are different for the boundary layer and for the bulk of the material, and different for the parallel and perpendicular directions of the dipoles, the "local" permittivity also becomes position-dependent and uniaxial:

$$\epsilon_t(z) = \epsilon_0 + \frac{N\alpha}{1 - N\alpha C_t(z)/\epsilon_0} \qquad (5.96)$$

$$\epsilon_n(z) = \epsilon_0 + \frac{N\alpha}{1 - N\alpha C_n(z)/\epsilon_0} \qquad (5.97)$$

Here, $C_t(z) = C_n(z) = 1/3$ for particles inside the layer and $C_t(z) = 0.346$, $C_n(z) = 0.307$ for the particles located near the layer surfaces. Axis z is orthogonal to the interface, α is the inclusion polarizability, and $N = 1/a^3$ is the inclusion concentration. The variation of the interaction constant is only important if the layer thickness is very small; otherwise, the contribution from a boundary layer whose thickness is of the order of the lattice period can be neglected, and the effective permittivity of the material is uniform and isotropic. To analyze the surface effect, we introduce the averaged material parameters, averaging across the layer thickness D:

$$\epsilon_D^t = \frac{1}{D} \int_0^D \epsilon_t(z)\, dz \qquad (5.98)$$

$$\epsilon_D^n = D \left(\int_0^D \frac{dz}{\epsilon_n(z)} \right)^{-1} \qquad (5.99)$$

Here, subindex D emphasizes that the averaging refers to the slab thickness and not to the cell volume as above. Equations (5.98) and (5.99) were derived in Chapter 2 from the assumption of the locally quasistatic distribution of the fields across the slab thickness. After calculation of the averaged effective parameters of the slab, the reflection and transmission coefficients can be solved as for a usual uniaxial slab.

    The two boundary layers occupy a volume fraction $2a/D$ of the layer, and the fraction of the internal volume is $1 - 2a/D$. Probably the easiest way to evaluate the integrals in (5.98) and (5.99) is to express the parameters in terms of the relative bulk permittivity ϵ_r. Using the Clausius-Mossotti formula (5.81)

$$\frac{N\alpha}{3\epsilon_0} = \frac{\epsilon_r - 1}{\epsilon_r + 2} \qquad (5.100)$$

we can write the relative local permittivity in the boundary layer as

$$\epsilon_{1\,t,n} = 1 + \frac{3(\epsilon_r - 1)/(\epsilon_r + 2)}{1 - 3C_{1\,t,n}(\epsilon_r - 1)/(\epsilon_r + 2)} \tag{5.101}$$

where $C_{1\,t,n}$ are the Lorentz factors in the boundary layer given in Table 5.1. Then, the averaged parameters of the slab become

$$\epsilon_D^t = \frac{2a}{D}\epsilon_{1t} + \left(1 - \frac{2a}{D}\right)\epsilon_r \tag{5.102}$$

$$\epsilon_D^n = \left[\frac{2a}{D}\frac{1}{\epsilon_{1n}} + \left(1 - \frac{2a}{D}\right)\frac{1}{\epsilon_r}\right]^{-1} \tag{5.103}$$

As a numerical example, for a slab of four layers of inclusions (that is, $2a/D = 0.5$) and the bulk relative permittivity value $\epsilon_r = 2$, we find that $\epsilon_D^t = 2.006$ and $\epsilon_D^n = 1.987$. For high permittivity values the effect is much more pronounced. For instance, for $\epsilon_r = 20$ we get $\epsilon_D^t = 23$ and $\epsilon_D^n = 16.2$.

In conclusion of our discussion of the Maxwell Garnett model, we note that this model has also been extended for mixtures of bianisotropic particles (see e.g., [11]). Other models of composites with negligible spatial dispersion exist [15, 16].

5.3 TRANSMISSION LINES WITH PERIODICAL INSERTIONS, PASSBANDS, AND STOPBANDS

The Maxwell Garnett model, as well as other mixing rules [15], can be used if the characteristic distances between the individual inclusions are small compared to the wavelength. This means that the spatial dispersion is negligible. At shorter wavelengths and higher frequencies other models are needed, which properly describe strong spatial dispersion effects. Most interesting features are found in periodical media, where strong spatial dispersion is due to resonant distances between inclusions. We start our studies of periodical media considering a simple example of a waveguide with periodically inserted inhomogeneities. This is also a model of various more complicated artificial media, because the transmission-line model (Chapter 2) can be applied to layers of various materials.

5.3.1 Dispersion Equation

Consider a transmission line with lumped periodical insertions. Let us denote the line wave impedance by η and suppose that every inclusion is modeled by a bulk reactive impedance Z_{load} (Figure 5.5). Let us find the dispersion relation (relation between the propagation factor along the transmission line q and the frequency) assuming that the distance between the insertions is large, meaning that the higher-order modes generated in the transmission line by every inhomogeneity decay at this distance. This means that the coupling between the inhomogeneities is only due to the fundamental mode fields.

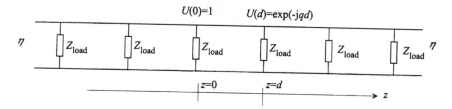

Figure 5.5 Transmission line (wave impedance η) with periodical bulk insertions.

Consider one reference inclusion and let the voltage at that position $U(0) = U|_{z=0} = 1$. The current through the load is then $U/Z_{\text{load}} = 1/Z_{\text{load}}$. This inhomogeneity scatters. First, we need to know the amplitude of the wave scattered by only one insertion in the matched regular line. The total load for this current source is $\eta/2$; thus, the amplitude of the scattered voltage is $U_{\text{sc}} = \eta/(2Z_{\text{load}})$. Now remember that we have an infinite number of such insertions and the field at the position of the reference inclusion ($z = 0$) is created by all the other inclusions:

$$U(0) = \frac{\eta}{2Z_{\text{load}}} \sum_{n=-\infty}^{\infty} e^{-jk|n|d} e^{-jqnd} = 1 \qquad (5.104)$$

Here, k is the propagation factor of the regular transmission line with no insertions. The infinite sum can be calculated exactly, assuming first that there are small losses in the background medium, so that convergence is secured, and then taking the limit $\text{Im}\{k\} \to 0$:

$$\sum_{n=-\infty}^{\infty} e^{-jk|n|d} e^{-jqnd} = 1 + \sum_{n=1}^{\infty} \exp[-j(q+k)nd] + \sum_{n=1}^{\infty} \exp[j(q-k)nd]$$

$$= 1 - \frac{\exp[-j(q+k)d]}{\exp[-j(q+k)d] - 1} - \frac{\exp[j(q-k)d]}{\exp[j(q-k)d] - 1} = j\frac{\sin kd}{\cos kd - \cos qd} \qquad (5.105)$$

The summation formula for geometrical progressions has been used. From this summation and (5.104), we find a relation between the propagation factor q and the wavenumber k, that is, the dispersion equation:

$$\cos qd = \cos kd + j\frac{\eta}{2Z_{\text{load}}} \sin kd \qquad (5.106)$$

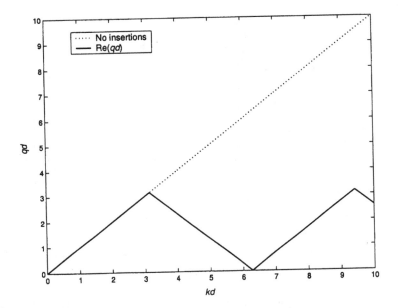

Figure 5.6 Dispersion curves for an unloaded transmission line, and the same line periodically loaded by very large bulk impedances. The properties of these two transmission lines are nearly identical, although the dispersion plots look very different.

Alternatively, the same result can be derived using the transmission matrix of a unit cell (see [31]). As a simple check, we observe that for $Z_{\text{load}} \to \infty$ (regular lines with no insertions) we have $\cos qd = \cos kd$; that is, $q = k$.

It is obvious that all solutions with $q \to q + (2\pi n)/d$, $n = 1, 2, 3, \ldots$ are equivalent, due to periodicity. In addition, in symmetrical systems such as the considered transmission line, $q \to -q$ is also a solution. For this reason, it is enough to consider solutions of (5.106) only in the interval $0 \le \text{Re}\{q\} \le \pi/d$. This is illustrated by Figure 5.6. Here we show the dispersion plot for an empty regular transmission line (which is obviously the straight line $q = k$), together with the solution of (5.106) for a very large value of the load impedance Z. For the periodical line, the values of qd are confined by the limits 0 and π, but for example in the frequency range $\pi < kd < 2\pi$, the plotted values of qd become the same as that for the empty line for $qd \to 2\pi - qd$. A similar concept is used in solid-state physics: Dispersion plots are reduced to the first Brillouin zone.

In the presence of loads, at some frequencies propagation along the line is impossible. This is obvious from (5.106): The right-hand side can take values larger than one (we assume purely reactive loads) within some frequency intervals. If the loads are capacitive, $Z_{\text{load}} = 1/(j\omega C_{\text{load}})$, the coefficient in

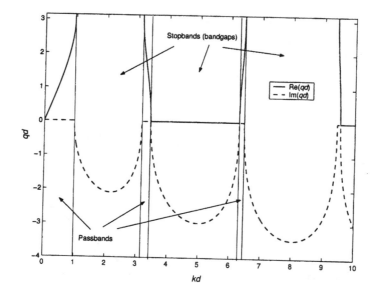

Figure 5.7 Typical dispersion curves for periodical capacitively loaded transmission lines.

(5.106) is of the form

$$j\frac{\eta}{2Z_{\text{load}}} = -kd\,A \tag{5.107}$$

where the constant $A = \eta C_{\text{load}}/(2\sqrt{\epsilon_0\mu_0}\,d)$. The curves plotted in Figure 5.7 have been calculated for $A = 2$. At low frequencies, where $kd \ll 1$, we can use the Taylor expansions for the cosine and sine functions in (5.106) to solve for q:

$$q \approx k\sqrt{1+A} \tag{5.108}$$

In terms of the per-unit-length parameters L and C of the transmission line, this reads [34]:

$$q \approx \omega\sqrt{L}\sqrt{C + \frac{C_{\text{load}}}{d}} \tag{5.109}$$

This means that the transmission line behaves as an effectively homogeneous line with the effective capacitance per unit length $C + C_{\text{load}}/d$: The total effective capacitance is increased by the load capacitance *averaged* over the period of insertions d. With increasing frequency, the line becomes dispersive, and eventually there appears a stopband (Figure 5.7).

For inductive loads $Z_{\text{load}} = j\omega L_{\text{load}}$, the coefficient in (5.106) takes the form

$$j\frac{\eta}{2Z_{\text{load}}} = \frac{A}{kd} \tag{5.110}$$

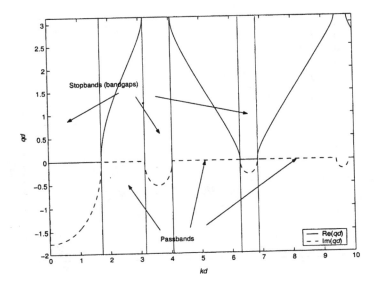

Figure 5.8 The same plots as in Figure 5.7 for inductive loads.

where the constant $A = \eta\sqrt{\epsilon_0\mu_0}\,d/(2L_{\text{load}})$. Plots in Figure 5.8 have been calculated for $A = 2$. In this case there is a low-frequency stopband, which is obviously due to very low impedances of loads at these frequencies that effectively create a short circuit of the transmission line.

In the design of metamaterials (Chapter 6), the most common situation is when the composite inclusions are resonant particles, sometimes electrically controllable. Resonances are needed to provide strong response to the incident field, especially in compact structures with not very many particles. With this in mind, let us continue to study the example of a periodically loaded transmission line and consider resonant loads.

First, we assume that each load can be modeled as a series connection of a capacitance and an inductance. Thus, a series resonance will take place, at which the load has a very low impedance. For the load impedance $Z_{\text{load}} = j\omega L_{\text{load}} + 1/(j\omega C_{\text{load}})$, the dispersion equation (5.106) is

$$\cos qd = \cos kd - \frac{\eta\omega C_{\text{load}}}{2(1 - \omega^2 L_{\text{load}} C_{\text{load}})}\sin kd = \cos kd - \frac{Akd}{B - (kd)^2}\sin kd \tag{5.111}$$

where constants A and B can be adjusted by choosing appropriate values of the load inductance and capacitance. In the example shown in Figure 5.9, $A = 0.1$ and $B = 0.09$. Near the resonance of the loads, a new stopband appears, whose position and width depend on the loads. The physical reason for this stopband is the fact that the load impedance is very small near the resonance, so the loads are highly reflective. For loads composed of a parallel

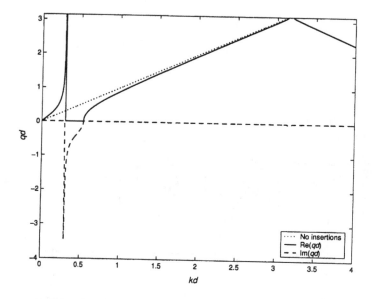

Figure 5.9 Dispersion curves for a periodical transmission line loaded by a resonant series connection of a capacitance and inductance.

connection of a capacitance and an inductance, we get

$$\cos qd = \cos kd + \frac{B - (kd)^2}{Akd} \sin kd \qquad (5.112)$$

The plot in Figure 5.10 is for $A = 1$ and $B = 0.09$. At the resonance frequency of the loads the propagation factor is not affected by the loads. At lower frequencies there is a stopband, since the load behaves as an inductance. At high frequencies the dispersion plot is similar to that for capacitively loaded lines.

5.3.2 Power Flow and Forward and Backward Waves

The dispersion relation for periodically loaded transmission lines was derived above considering only voltages and currents at the load positions. Let us next look into the voltage (field) distribution inside the cells, between the loads.[10] Consider one period of an infinite structure, shown in Figure 5.11. Because we have assumed that the higher-order modes decay at distance d, the voltage in this transmission-line section is the sum of two fundamental-mode waves traveling in the opposite directions:

$$U(z) = Ae^{-jkz} + Be^{jkz} \qquad (5.113)$$

---

[10]The following derivations in this section were made together with S.I. Maslovski.

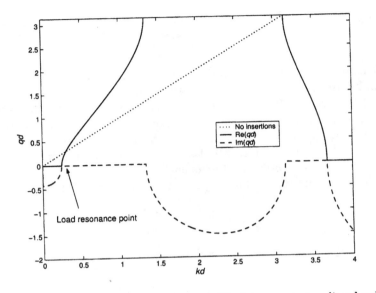

Figure 5.10 Dispersion curves for a periodical transmission line loaded by a resonant parallel connection of a capacitance and inductance.

where k is the propagation factor in the unloaded line, and A, B are the amplitude coefficients. The Floquet theorem reveals that

$$U(d) = U(0)e^{-jqd} \quad \Rightarrow \quad (A+B)e^{-jqd} = Ae^{-jkd} + Be^{jkd} \quad (5.114)$$

From here we can find the reflection coefficient at $z = 0$:

$$R(0) = \frac{B}{A} = -\frac{e^{-jqd} - e^{-jkd}}{e^{-jqd} - e^{+jkd}} = e^{-jkd}\frac{\sin[(k-q)d/2]}{\sin[(k+q)d/2]} \quad (5.115)$$

As in any transmission line, the reflection coefficient changes along the line as $R(z) = R(0)e^{2jkz}$; thus, at the cell center $z = d/2$ we have a simpler

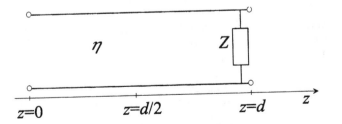

Figure 5.11 One period of a transmission line with periodically inserted loads.

expression

$$R(d/2) = R = \frac{\sin[(k-q)d/2]}{\sin[(k+q)d/2]} \qquad (5.116)$$

This formula is a good approximation for the reflection coefficient from a semi-infinite line or a half-space filled by an electromagnetic crystal. We learned in Section 5.2.3 that the local field differs from its bulk value only in a few cells nearest to an interface. Neglecting the effect of this very thin transition layer, the reflection coefficient from an interface is the same as that in the bulk. Of course, the other restriction is a possibility to neglect higher-order modes between inhomogeneities.

The impedance in the line at point $z = d/2$ (Figure 5.11) is

$$Z = \eta \frac{1+R}{1-R} = \eta \frac{\tan(kd/2)}{\tan(qd/2)} \qquad (5.117)$$

Finally, the power flow along the periodically loaded transmission line is given by

$$P = \frac{1}{2}\,\mathrm{Re}\{UI^*\} = \frac{1}{2}\,\mathrm{Re}\{Z\}|I|^2 \qquad (5.118)$$

For lossless lines, this value is of course independent from coordinate z. The direction of the power flow is thus defined by the sign of the real part of the line impedance (5.117).

Effective Parameters

In the low-frequency regime, when $kd \ll 1$ and $qd \ll 1$, we can approximate (5.116) and (5.117) as

$$R \approx \frac{k-q}{k+q}, \qquad Z \approx \eta \frac{k}{q} \qquad (5.119)$$

This tells us that in this situation an effective permittivity can be introduced by writing $q = \omega\sqrt{\epsilon_{\mathrm{eff}}\mu_0}$, which gives

$$\epsilon_{\mathrm{eff}} = \epsilon_0 \frac{q^2}{k^2} \qquad (5.120)$$

Here, μ_0 and ϵ_0 are the parameters of the medium filling the transmission line. Indeed, substituting this together with $k = \omega\sqrt{\epsilon_0\mu_0}$ into (5.119), we are left with the familiar formulas

$$R = \frac{\sqrt{\epsilon_0} - \sqrt{\epsilon_{\mathrm{eff}}}}{\sqrt{\epsilon_0} + \sqrt{\epsilon_{\mathrm{eff}}}}, \qquad Z = \sqrt{\frac{\mu_0}{\epsilon_{\mathrm{eff}}}} \qquad (5.121)$$

for a homogeneous material with the parameters ϵ_{eff} and μ_0. It is interesting to note that the same result for the effective parameters follows from averaging the microscopic voltage and current (microscopic fields, in the modeling of materials) over one period. Let us introduce the averaged voltage (or electric field) as

$$\widehat{U} = \frac{1}{d} \int_0^d U(z)\, dz = \frac{1}{d} \int_0^d A(e^{-jkz} + Re^{jkz})\, dz \qquad (5.122)$$

and current (or magnetic field) as

$$\widehat{I} = \frac{1}{d} \int_0^d I(z)\, dz = \frac{1}{\eta d} \int_0^d A(e^{-jkz} - Re^{jkz})\, dz \qquad (5.123)$$

where R is given by (5.116). If we now *define* the wave impedance in the effective medium as $Z = \widehat{U}/\widehat{I}$ and substitute (5.122) and (5.123), the result is

$$Z = \eta \frac{1 - e^{-jkd} - R(1 - e^{jkd})}{1 - e^{-jkd} + R(1 - e^{jkd})} \qquad (5.124)$$

For small kd and qd, this simplifies to the same result as before:

$$Z \approx \eta \frac{1 + R}{1 - R} \approx \eta \frac{k}{q} \qquad (5.125)$$

However, if the period is not small compared to the wavelength, the impedances for the microscopic fields (5.117) and for the averaged fields (5.124) are different.

This fact is important for the problem of introducing effective material parameters. Sometimes the effective parameters of composite materials are introduced, demanding the reflection coefficient for normally incident plane waves be given by the simple formula known for *homogeneous magnetodielectrics*, if the effective material parameters are substituted. Because the wave impedance that defines the reflection coefficient (5.117) and the wave impedance that connects the macroscopic (averaged) fields in the medium (5.124) are in general different, the effective permittivity introduced via the reflection coefficient can be in fact used only to calculate the reflection coefficient at the normal incidence. This effective permittivity is *not* equal to the coefficient in the material relation $\mathbf{D} = \epsilon_{\text{eff}} \mathbf{E}$. For example, the reflection at oblique incidence is not given by the standard formula for the reflection from a homogeneous interface with the effective medium. In other words, this effective permittivity has to be considered as a function of the incidence angle, so the very meaning of the effective permittivity is lost. Only if the spatial dispersion can be neglected ($kd \ll 1$, $qd \ll 1$), then the wave impedances (5.117) and (5.125) coincide, and the effective permittivity introduced via the reflection coefficient has the full physical meaning of the effective medium parameter. The reason for the problem is that spatially dispersive media cannot be described in terms of only effective permittivity. If the spatial dispersion effects are weak but not negligible, the constitutive relations become more complicated [11]. If the spatial dispersion is strong (in our example, distances between loads comparable to the wavelength), the effective medium model cannot be used at all.

Forward and Backward Waves

With the simple results for the loaded transmission line we are ready to discuss the power flow direction, the phase velocity, and so-called *backward waves*. In backward waves the directions of the phase velocity and the power flow are the opposite. For example, artificial media with negative real parts of the permittivity and permeability support backward waves, and it is this property that defines the most interesting phenomena in these materials. More on this subject can be found in Chapter 6. Here we consider direct and backward waves in generic periodic media. Let us start from the simplest case of a periodically loaded transmission line with very large load impedances, which means that the properties of the line are nearly the same as that of an empty transmission line. The dispersion plot for this case is shown in Figure 5.6. Calculation of the wave impedance (5.117) shows that $Z/\eta \approx 1$ in the regions where q grows with k ($0 < kd < \pi$, $2\pi < kd < 3\pi$, and so forth) and $Z/\eta \approx -1$ elsewhere ($\pi < kd < 2\pi$, $3\pi < kd < 4\pi$, and so forth). From (5.118), we conclude that the power flow is in the positive direction of z in the regions $0 < kd < \pi$, $2\pi < kd < 3\pi$... and in the negative direction in $\pi < kd < 2\pi$, $3\pi < kd < 4\pi$.... This appears to be in accord with the fact that the group velocity as seen from the plot in Figure 5.6 is also negative in the same regions where the power flows in the negative direction. However, in the same plot both q and k are positive everywhere, so it *seems* that the phase velocity is always positive, and we can conclude that in the regions of the negative power flow direction, the line supports only backward waves. This can be quite confusing because on the other hand the line is in fact nearly regular, since with the load impedance values tending to infinity, the loads cannot influence the line properties. Obviously, the waves are always forward waves in ordinary regular transmission lines (as well as in free space, for example).

The reason for the confusion can be easily understood if we look at the reflection coefficient in the line (5.116). As is clear from (5.117), in the regions of "direct" waves, where $Z/\eta \approx 1$, the reflection coefficient is nearly zero, which means that the voltage is simply $U(z) = Ae^{-jkz}$. But in the "backward" wave regions, where $Z/\eta \approx -1$, the reflection coefficient tends to infinity, thus $A = 0$, and $U(z) = Be^{jkz}$. This means that in these regions the phase between the loads also advances in the negative direction, so these regions of the plot simply show normal forward waves but traveling in the negative direction of the z axis. Why then does the plot look like there are

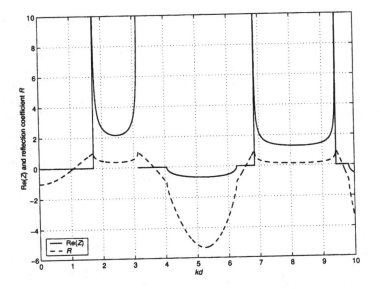

Figure 5.12 The real part of impedance (5.117) and the reflection coefficient (5.116) for the inductively loaded transmission line of Figure 5.8.

backward waves in every second passband? This is because the reduction to the first Brillouin zone has been done considering only voltages at the positions of loads, or, in more general terms, only at discrete points in space separated by a period. If we look at how the phase of the currents in the loads advances from one inclusion to the next, it will indeed appear as a phase wave moving to the positive direction for all kd (as in Figure 5.6). But the phase velocity of the electromagnetic field in space between the inclusions can be different, as we have just demonstrated. We conclude this discussion stressing that in defining forward and backward waves care should be taken in defining the phase velocity in the medium.

Figure 5.12 provides an illustration of this concept for a capacitively loaded line. In the first passband the power flow is in the positive direction ($\mathrm{Re}\{Z\} > 0$), and the reflection coefficient is smaller than unity, so the forward-directed wave dominates. This is a usual forward wave. In the second passband the power flow direction is reversed, but the reflection coefficient is larger than unity in the absolute value. The phase velocity of the microscopic field is also reversed, and we conclude that in this sense it is also a forward wave. The wave of currents in the loads $I(nd) = I(0)e^{-jqnd}$, however, is a backward wave in this interval. Note that if the formula for the reflection coefficient is used to calculate reflection from a semi-infinite structure, the sign of the propagation constant should be chosen properly, so that the wave power propagates in the direction from the source. In this example, inside the second passband we should replace $qd \to 2\pi - qd$, which is equivalent to changing the sign of q.

A clear example is a wave along a spiral made of a conducting wire. Spiral structures are used as phase delay systems in microwave tubes, and it is known that they support backward waves (utilized in backward-wave microwave oscillators). However, it is obvious that the phase velocity of an electromagnetic wave traveling *along the wire* of a spiral is positive in the sense that the phase advances in the same direction as that of the energy flow. If the spiral diameter and pitch are large compared to the wire diameter, and the wire is a good conductor, this wave is a usual electromagnetic wave traveling with the speed of light along the conducting wire. But, if we are only interested in the phase of the field at the spiral axis and consider how the phase of this field changes from one turn to the other, we find that the phase velocity is very different from the speed of light, and can even be negative. In this sense the phase velocity is not uniquely defined. Which definition of the phase velocity to use depends on the application. In microwave tubes, an electron beam propagates along the spiral axis and interacts with the delayed electromagnetic wave. The velocity of the wave along the wire is of little relevance. On the other hand, if a spiral conductor is used to introduce a phase shift, the phase velocity of the wave traveling along the wire is more important.

Our last comment here is about modeling of materials by effective parameters. We saw that the phase velocity and wave impedance are different for the microscopic and the averaged fields. Moreover, different results for the phase velocity will follow if different averaging methods are used.

5.4   ONE-DIMENSIONAL ELECTROMAGNETIC CRYSTALS

Next, we study electromagnetic properties of one-dimensional electromagnetic crystals, or *photonic bandgap structures*, composed of periodically arranged plane layers of isotropic dielectrics or magnetodielectrics. This system allows analytical solutions, and its study is helpful in understanding many important features also found in more complex systems.

5.4.1   Dispersion Equation and Low-Frequency Model

For simplicity, we will write formulas for structures whose period contains two layers (thicknesses $d_{1,2}$, permittivities $\epsilon_{1,2}$, and permeabilities $\mu_{1,2}$). The period of the "crystal" is then $d = d_1 + d_2$ (Figure 5.13). Extension to the general case of several layers comprising every period is straightforward.

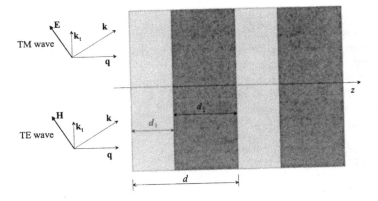

Figure 5.13 Geometry of a one-dimensional electromagnetic crystal. Only two periods are shown. The slabs are infinite in the plane orthogonal to z.

Eigenvalue Equation

The eigenvalue equation can be conveniently derived using the transmission-matrix model of slabs (Section 2.4). Let us write the linear relations between the tangential electric and magnetic fields at the two sides of a plane material layer in matrix form, as in (2.111):

$$\left(\begin{array}{c} \mathbf{E}_{t+} \\ \mathbf{n} \times \mathbf{H}_{t+} \end{array} \right) = \left(\begin{array}{cc} \overline{\overline{a}}_{11} & \overline{\overline{a}}_{12} \\ \overline{\overline{a}}_{21} & \overline{\overline{a}}_{22} \end{array} \right) \cdot \left(\begin{array}{c} \mathbf{E}_{t-} \\ \mathbf{n} \times \mathbf{H}_{t-} \end{array} \right) \tag{5.126}$$

The dyadic coefficients for isotropic slabs are given by (2.112) and (2.113). Each period of the analyzed structure contains two layers, and the transmission matrix for one period is the product of the transmission matrices of the two layers. Denoting the components of this product by $\overline{\overline{A}}_{ij}$, we can write

$$\left(\begin{array}{c} \mathbf{E}_{t+} \\ \mathbf{n} \times \mathbf{H}_{t+} \end{array} \right) = \left(\begin{array}{cc} \overline{\overline{A}}_{11} & \overline{\overline{A}}_{12} \\ \overline{\overline{A}}_{21} & \overline{\overline{A}}_{22} \end{array} \right) \cdot \left(\begin{array}{c} \mathbf{E}_{t-} \\ \mathbf{n} \times \mathbf{H}_{t-} \end{array} \right) \tag{5.127}$$

Looking for the eigenwaves, we can make use of the Floquet theorem (4.2), which tells us that the fields at a distance of one period differ by a phase factor $e^{-jq(d_1+d_2)} = e^{-jqd}$. Combining with (5.127), we have

$$e^{-jqd} \left(\begin{array}{c} \mathbf{E}_{t-} \\ \mathbf{n} \times \mathbf{H}_{t-} \end{array} \right) = \left(\begin{array}{cc} \overline{\overline{A}}_{11} & \overline{\overline{A}}_{12} \\ \overline{\overline{A}}_{21} & \overline{\overline{A}}_{22} \end{array} \right) \cdot \left(\begin{array}{c} \mathbf{E}_{t-} \\ \mathbf{n} \times \mathbf{H}_{t-} \end{array} \right) \tag{5.128}$$

Thus, to find the propagation constants of eigenwaves q, corresponding to nontrivial solutions for the fields, we should equate the determinant of the matrix

$$\left(\begin{array}{cc} \overline{\overline{A}}_{11} - e^{-jqd}\overline{\overline{I}}_t & \overline{\overline{A}}_{12} \\ \overline{\overline{A}}_{21} & \overline{\overline{A}}_{22} - e^{-jqd}\overline{\overline{I}}_t \end{array} \right) \tag{5.129}$$

to zero.

It is easier to make actual calculations separately for TE and TM polarized modes, using the basis of eigenvectors of the transmission matrix dyadic coefficients for isotropic slabs (Section 2.4). For TM fields, the coefficients of the transmission matrices for individual slabs are scalars given by (2.118) and (2.120):

$$\overline{\overline{A}}^{TM} = \begin{pmatrix} \cos(\beta_1 d_1) & j\frac{\eta_1\beta_1}{k_1}\sin(\beta_1 d_1) \\ j\frac{k_1}{\eta_1\beta_1}\sin(\beta_1 d_1) & \cos(\beta_1 d_1) \end{pmatrix}$$

$$\cdot \begin{pmatrix} \cos(\beta_2 d_2) & j\frac{\eta_2\beta_2}{k_2}\sin(\beta_2 d_2) \\ j\frac{k_2}{\eta_2\beta_2}\sin(\beta_2 d_2) & \cos(\beta_2 d_2) \end{pmatrix} = \begin{pmatrix} A_{11} & A_{12} \\ A_{21} & A_{22} \end{pmatrix} \quad (5.130)$$

and the coefficients of the transmission matrix over one period A_{ij} are scalars. Here we use the same notations as in Chapter 2: $\beta_{1,2} = \sqrt{k_{1,2}^2 - k_t^2}$ are the propagation factors along the z-axis. Calculating the appropriate determinant (5.129), we arrive at equation

$$A_{11}A_{22} - A_{12}A_{21} - e^{-jqd}(A_{11} + A_{22}) + e^{-2jqd} = 0 \quad (5.131)$$

From the perspective of substituting the expressions for A_{ij} this looks complicated, but in fact it is not, because the determinants of the transmission matrices for both layers equal unity, as is obvious from (5.130). Thus, also $A_{11}A_{22} - A_{12}A_{21} = 1$, and the eigenvalue equation (5.131) simplifies as

$$\cos(qd) = \frac{1}{2}(A_{11} + A_{22}) = \frac{1}{2}\text{trace}\{\overline{\overline{A}}\} \quad (5.132)$$

Finally, a simple calculation leads to the equation for the propagation factor q:

$$\cos(qd) = \cos(\beta_1 d_1)\cos(\beta_2 d_2) - \frac{1}{2}\left(\frac{\eta_1\beta_1 k_2}{\eta_2\beta_2 k_1} + \frac{\eta_2\beta_2 k_1}{\eta_1\beta_1 k_2}\right)\sin(\beta_1 d_1)\sin(\beta_2 d_2)$$

$$(5.133)$$

The eigenvalue equation for TE modes can be obtained from (5.133), replacing $\eta_{1,2}\beta_{1,2}/k_{1,2}$ by $\eta_{1,2}k_{1,2}/\beta_{1,2}$, as seen from (2.118) and (2.120). For the propagation in the direction orthogonal to the interfaces between layers, we have $\beta_{1,2} = k_{1,2}$, and (5.133) reduces to

$$\cos(qd) = \cos(k_1 d_1)\cos(k_2 d_2) - \frac{1}{2}\frac{\epsilon_2\mu_1 + \epsilon_1\mu_2}{\sqrt{\epsilon_1\mu_1\epsilon_2\mu_2}}\sin(k_1 d_1)\sin(k_2 d_2) \quad (5.134)$$

Effective Material Parameters

Let us consider if it is possible to introduce effective averaged parameters of the structure in the low-frequency regime, when both layers are thin compared with the wavelength: $k_1 d_1 \ll 1$, $k_2 d_2 \ll 1$, and also $qd \ll 1$. Here we will consider only dielectric slabs, assuming $\mu_1 = \mu_2 = \mu_0$. Since the period is small in wavelengths, we can expect that the material can be modeled by a uniaxial permittivity dyadic

$$\overline{\overline{\epsilon}}_{\text{eff}} = \epsilon_t \overline{\overline{I}}_t + \epsilon_z \mathbf{z}_0 \mathbf{z}_0 \tag{5.135}$$

The medium properties are clearly different for electric fields polarized along the z-axis of the structure or in the plane of the interfaces (Figure 5.13). The dispersion relation for plane waves in uniaxial dielectrics is well known (e.g., [35]):

$$q^2 + k_t^2 = \omega^2 \mu_0 \epsilon_t, \qquad \text{TE modes} \tag{5.136}$$

$$q^2 + \frac{\epsilon_t}{\epsilon_z} k_t^2 = \omega^2 \mu_0 \epsilon_t, \qquad \text{TM modes} \tag{5.137}$$

Starting from TM modes, we expand the trigonometric functions in (5.133) in the Taylor series:

$$(qd)^2 = (\beta_1 d_1)^2 + (\beta_2 d_2)^2 + \left(\frac{\eta_1 \beta_1 k_2}{\eta_2 \beta_2 k_1} + \frac{\eta_2 \beta_2 k_1}{\eta_1 \beta_1 k_2} \right) \beta_1 d_1 \beta_2 d_2 \tag{5.138}$$

Now we can try to rewrite this in the form of (5.137) and identify the equivalent material parameters. After simple algebra (recall that $\beta_{1,2}^2 = k_{1,2}^2 - k_t^2$), (5.138) reduces to

$$q^2 + \frac{(\epsilon_1 d_1 + \epsilon_2 d_2)(\epsilon_1 d_2 + \epsilon_2 d_1)}{\epsilon_1 \epsilon_2 (d_1 + d_2)^2} k_t^2 = \omega^2 \mu_0 \frac{\epsilon_1 d_1 + \epsilon_2 d_2}{d_1 + d_2} \tag{5.139}$$

Expanding the dispersion equation for TE modes, we get

$$q^2 d^2 = \beta_1^2 (d_1^2 + d_1 d_2) + \beta_2^2 (d_2^2 + d_1 d_2) \tag{5.140}$$

that is equivalent to

$$q^2 + k_t^2 = \omega^2 \mu_0 \frac{\epsilon_1 d_1 + \epsilon_2 d_2}{d_1 + d_2} \tag{5.141}$$

Comparing (5.139) and (5.141) with the dispersion relations for uniaxial dielectrics (5.136) and (5.137), we conclude that one-dimensional electromagnetic crystals indeed behave at low frequencies as effective uniaxial dielectrics, and identify the effective parameters:

$$\epsilon_t = \frac{\epsilon_1 d_1 + \epsilon_2 d_2}{d_1 + d_2} \tag{5.142}$$

$$\epsilon_z = \frac{d_1 + d_2}{\frac{d_1}{\epsilon_1} + \frac{d_2}{\epsilon_2}} \tag{5.143}$$

Note that (5.142) coincides with the result of the locally quasistatic averaging of the tangential electric field over one period; see Chapter 2, formula (2.99). The other component, ϵ_z, can also be found from a simple special case of propagation along the interfaces, when the wave does not feel the periodicity of the medium. If the electric field is orthogonal to the interfaces, the locally quasistatic averaging leads to (2.83), or, for our two-layer period, to formula (5.143). However, we stress that consideration of such special propagation directions can be used only as a check of the general result. Only from the general consideration above could we prove that the material is an effective uniaxial dielectric.

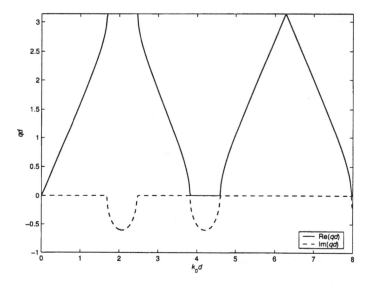

Figure 5.14 Dispersion plot for a one-dimensional electromagnetic crystal formed by dielectric layers. In this example $d_2 = d_1$ and $\sqrt{\epsilon_2} = 2\sqrt{\epsilon_1}$.

Next we consider broadband behavior of the crystal and show a typical dependence of the normalized propagation factor as a function of the normalized frequency (axial propagation, $k_t = 0$; see Figure 5.14). Note that in the region of small $k_0 d$ the function is approximately linear. In this region the effective dielectric model (5.142) is valid.

5.4.2   Reflection and Transmission in Finite Structures

Reflection and transmission coefficients from one-dimensional crystals containing a finite number of periods (Figure 5.15) can be found using the vector-circuit model of Chapter 2. This technique was developed for chiral multilayers in [36, 37]. Let us denote by $\mathbf{E}_{t+}$ and $\mathbf{H}_{t+}$ the tangential fields

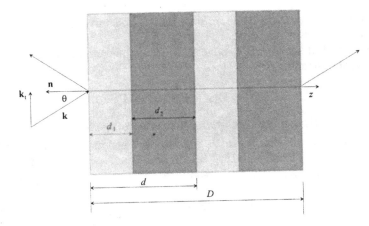

Figure 5.15 Excitation of a layered structure with a finite thickness D by a plane wave of an arbitrary polarization. Homogeneous media at the two sides of the structure can be different.

at the first interface from the source side (the left-hand side on Figure 5.15). $\mathbf{E}_{t-}$ and $\mathbf{H}_{t-}$ denote the tangential fields at the opposite side of the whole structure. We start from calculation of the transmission matrix $\overline{\overline{A}}$ with dyadic components $\overline{\overline{A}}_{ij}$ for the whole multilayer structure that connects these four field vectors:

$$\begin{pmatrix} \mathbf{E}_{t+} \\ \mathbf{n} \times \mathbf{H}_{t+} \end{pmatrix} = \begin{pmatrix} \overline{\overline{A}}_{11} & \overline{\overline{A}}_{12} \\ \overline{\overline{A}}_{21} & \overline{\overline{A}}_{22} \end{pmatrix} \cdot \begin{pmatrix} \mathbf{E}_{t-} \\ \mathbf{n} \times \mathbf{H}_{t-} \end{pmatrix} \tag{5.144}$$

[compare with (5.127)]. This transmission matrix is the product of matrices for individual layers that we know from Chapter 2, (2.111). The uniform source-free half-space on the right from the structure can be modeled by an equivalent dyadic bulk impedance $\overline{\overline{Z}}_0'$ given by (3.5). Substituting

$$\mathbf{E}_{t-} = \overline{\overline{Z}}_0' \cdot (\mathbf{n} \times \mathbf{H}_{t-}) \tag{5.145}$$

into (5.144), we can eliminate the fields on the right-hand side and find an equivalent dyadic impedance seen from the source side at the first interface:

$$\mathbf{E}_{t+} = \overline{\overline{Z}}_{\mathrm{eq}} \cdot (\mathbf{n} \times \mathbf{H}_{t+}) \tag{5.146}$$

where

$$\overline{\overline{Z}}_{\mathrm{eq}} = (\overline{\overline{A}}_{11} \cdot \overline{\overline{Z}}_0' + \overline{\overline{A}}_{12}) \cdot (\overline{\overline{A}}_{21} \cdot \overline{\overline{Z}}_0' + \overline{\overline{A}}_{22})^{-1} \tag{5.147}$$

If the structure is on an ideally conducting surface, then $\overline{\overline{Z}}_0' = 0$ and the last relation reads

$$\overline{\overline{Z}}_{\mathrm{eq}} = \overline{\overline{A}}_{12} \cdot \overline{\overline{A}}_{22}^{-1} \tag{5.148}$$

which is the same as the internal impedance of the equivalent Thévenin circuit (Section 2.4) for the structure.

The reflected tangential electric field $\mathbf{E}_t^{\text{ref}}$ depends on the tangential component of the incident field $\mathbf{E}_t^{\text{inc}}$ as

$$\mathbf{E}_t^{\text{ref}} = \mathbf{E}_{t+} - \mathbf{E}_t^{\text{inc}} = \overline{\overline{R}} \cdot \mathbf{E}_t^{\text{inc}} \tag{5.149}$$

We emphasize that in this relation there are tangential field components, and not the total fields. Dyadic $\overline{\overline{R}}$ is a two-dimensional dyadic operating in the plane tangential to the interfaces. If the dyadic wave impedance of the half-space where the source is located is $\overline{\overline{Z}}_0$, the reflection dyadic reads

$$\overline{\overline{R}} = \left(\overline{\overline{Z}}_{\text{eq}} - \overline{\overline{Z}}_0\right) \cdot \left(\overline{\overline{Z}}_{\text{eq}} + \overline{\overline{Z}}_0\right)^{-1} \tag{5.150}$$

The transmission dyadic, defined through $\mathbf{E}_{t-} = \overline{\overline{T}} \cdot \mathbf{E}_t^{\text{inc}}$, can be found in form [36, 37]

$$\overline{\overline{T}} = \left(\overline{\overline{A}}_{11} + \overline{\overline{A}}_{12} \cdot \overline{\overline{Z}}_0'^{-1}\right)^{-1} \cdot \left(\overline{\overline{I}}_t + \overline{\overline{R}}\right) \tag{5.151}$$

For linearly polarized incident waves, it is convenient to divide the dyadics into parts according to the polarization state and write

$$\overline{\overline{R}} = R^{TE} \frac{\mathbf{k}_t \times \mathbf{n}\,\mathbf{k}_t \times \mathbf{n}}{k_t^2} + R^{TM} \frac{\mathbf{k}_t \mathbf{k}_t}{k_t^2} + R^{EM} \frac{\mathbf{k}_t\,\mathbf{k}_t \times \mathbf{n}}{k_t^2} + R^{ME} \frac{\mathbf{n} \times \mathbf{k}_t\,\mathbf{k}_t}{k_t^2} \tag{5.152}$$

and

$$\overline{\overline{T}} = T^{TE} \frac{\mathbf{k}_t \times \mathbf{n}\,\mathbf{k}_t \times \mathbf{n}}{k_t^2} + T^{TM} \frac{\mathbf{k}_t \mathbf{k}_t}{k_t^2} + T^{EM} \frac{\mathbf{k}_t\,\mathbf{k}_t \times \mathbf{n}}{k_t^2} + T^{ME} \frac{\mathbf{n} \times \mathbf{k}_t\,\mathbf{k}_t}{k_t^2} \tag{5.153}$$

These formulas are written for the general case when the layers can be anisotropic or chiral, in which case the linear polarization is not preserved in reflection and transmission. The superscripts TM and TE refer to copolarized reflection or transmission of the incident TM and TE plane waves, respectively. The superscripts EM and ME mark the cross-polarized reflection and transmission coefficients: EM for coupling from TE into TM wave and ME for coupling from TM into TE wave. For isotropic structures with no mode coupling, the cross-components in (5.152) and (5.153) equal zero.

The total reflected and transmitted fields for a TE-polarized incident field are

$$\mathbf{E}^{\text{ref}} = \left[R^{TE} \frac{\mathbf{k}_t \times \mathbf{n}}{k_t} + R^{EM} \left(\frac{\mathbf{k}_t}{k_t} - \mathbf{n}\tan\theta\right)\right] E^{\text{inc}} \tag{5.154}$$

$$\mathbf{E}^{\text{trans}} = \left[T^{TE} \frac{\mathbf{k}_t \times \mathbf{n}}{k_t} + T^{EM} \left(\frac{\mathbf{k}_t}{k_t} - \mathbf{n}\tan\theta\right)\right] E^{\text{inc}} \tag{5.155}$$

where E^{inc} is the complex amplitude of the electric field in the incident wave and θ stands for the angle of incidence ($\theta = 0$ corresponds to the normal incidence, Figure 5.15). As is seen, the copolarized reflection coefficient of the total field is the same as for the transverse field and it equals R^{TE}. The second term in (5.154) corresponds to the coupling from TE into TM waves. In terms of the total fields, the absolute value of the cross-polarized reflection coefficient is $|R^{EM}|\sqrt{1 + \tan^2 \theta}$.

For a TM-polarized incident wave, the total reflected and transmitted fields are

$$\mathbf{E}^{ref} = \left[R^{TM} \left(\frac{\mathbf{k}_t}{k_t} \cos\theta - \mathbf{n}\sin\theta \right) + R^{ME} \frac{\mathbf{n} \times \mathbf{k}_t}{k_t} \cos\theta \right] E^{inc} \qquad (5.156)$$

$$\mathbf{E}^{trans} = \left[T^{TM} \left(\frac{\mathbf{k}_t}{k_t} \cos\theta - \mathbf{n}\sin\theta \right) + T^{ME} \frac{\mathbf{n} \times \mathbf{k}_t}{k_t} \cos\theta \right] E^{inc} \qquad (5.157)$$

Now, the absolute value of the copolarized reflection coefficient is $|R^{TM}|$ and the cross-polarized reflected field is proportional to $R^{ME} \cos\theta$.

For lossless media the following energy conservation relations are true:

$$|R^{TM}|^2 + |T^{TM}|^2 + |\cos(\theta)R^{ME}|^2 + |\cos(\theta)T^{ME}|^2 = 1 \qquad (5.158)$$

$$|R^{TE}|^2 + |T^{TE}|^2 + \left|\frac{R^{EM}}{\cos(\theta)}\right|^2 + \left|\frac{T^{EM}}{\cos(\theta)}\right|^2 = 1 \qquad (5.159)$$

Furthermore, for reciprocal materials we have

$$R^{EM} = R^{ME}\cos(\theta), \qquad T^{EM} = T^{ME}\cos(\theta) \qquad (5.160)$$

so that the coupling between the orthogonally polarized fields is the same from TE to TM states and vice versa.

Figure 5.16 shows a calculated example[11] of transmission coefficient for a one-dimensional electromagnetic crystal (normal plane wave incidence) for various numbers of layers. Each period of the structure is made up of two isotropic layers. The parameters of the layers are the following: $\epsilon_1 = 2\epsilon_0$, $\epsilon_2 = 4\epsilon_0$, $\mu_1 = \mu_2 = \mu_0$. The electric thicknesses of the two layers are the same. At a certain frequency the electrical thicknesses are equal to the quarter of the free-space wavelength λ_0:

$$d_1\sqrt{\frac{\epsilon_1}{\epsilon_0}} = d_2\sqrt{\frac{\epsilon_2}{\epsilon_0}} = \frac{\lambda_0}{4} \qquad (5.161)$$

We denote the wavenumber at this frequency as $k_{\lambda/4}$. The wavenumber in Figure 5.16 is normalized to this value. From this graph, we see how the stopbands are formed when the number of layers is increased. Usually the bandgaps are well pronounced for layers having about five or more periods.

---

[11]Calculations have been made by I.S. Nefedov.

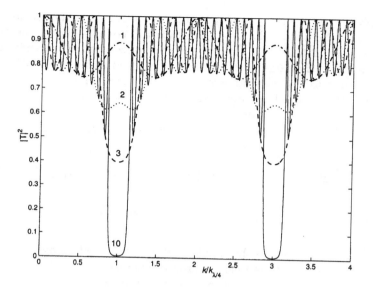

Figure 5.16 Power transmission coefficient through an electromagnetic crystal for different numbers of periods. The numbers of periods are shown near the corresponding curves.

5.5 WIRE MEDIA

As a very illuminating example of more complicated electromagnetic crystals, let us next consider artificial materials formed by lattices of thin infinite metal cylinders. In this chapter, we develop analytical models for this system [38] that will be used in the next chapter in the analysis of metamaterials and artificial impedance surfaces. This structure is known in microwave engineering as an artificial dielectric [39, 40], and here we will call it *wire medium*. Various numerical techniques suitable for the analysis of this structure have been published in [41–49], and some averaging methods in [50, 51].

We consider rectangular grids of infinite wires as depicted in Figure 5.17. The elementary cells have dimensions $\mathbf{a} \times \mathbf{b}$. The radius of wires is $r_0 \ll a, b$. We will demonstrate how to build the dispersion theory of this media and investigate the reflection of plane electromagnetic waves from a half-space filled by this artificial material. It will be also very important to analyze possible descriptions of the material in terms of the effective permittivity parameter in the quasistatic limit. Let us choose a coordinate system such that the z-axis is the axis of one (reference) wire, and the x- and y-axes are parallel to $\mathbf{a}$ and $\mathbf{b}$, respectively. In this coordinate system, the radius vectors of distances from the reference wire to the wire with numbers m, n can be written as $\mathbf{R}_{m,n} = m\mathbf{a} + n\mathbf{b}$. We assume that the wires are thin compared to

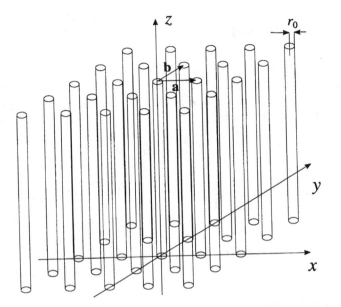

Figure 5.17 The geometry of wire media: a regular array of parallel ideally conducting thin wires.

the grid periods, so that polarization perpendicular to the wire axis is very small compared to the longitudinal one. Single layers forming this material were considered in Chapter 4.

In studies of eigenwaves in the infinite periodic structure, we assume the plane wave dependence of the current amplitudes in the form

$$\mathbf{I}_{m,n} = \mathbf{I}e^{-j(q_x am + q_y bn + q_z z)} \tag{5.162}$$

The field of every wire can be described using the Green function (e.g., [52])

$$G(r, z) = \frac{j}{4} H_0^{(2)}\left(\sqrt{k^2 - q_z^2}\, r\right) e^{-jq_z z} \tag{5.163}$$

where distance $r = |\mathbf{r}|$ is measured from the wire axis, and k is the wavenumber in the isotropic medium surrounding the wires. We need only the expression for the longitudinal electric field component, which can be obtained from (5.163) in the following form:

$$\mathbf{E}(r, z) = -\frac{\eta(k^2 - q_z^2)}{4k} H_0^{(2)}\left(\sqrt{k^2 - q_z^2}\, r\right) \mathbf{I}e^{-jq_z z} \tag{5.164}$$

where η is the wave impedance in the matrix material.

We write the linear relation between the induced current amplitude I and the longitudinal component of the local electric field E_z^{loc} in terms of susceptibility α:

$$I = \alpha E_z^{loc} \tag{5.165}$$

The susceptibility of a wire excited by a local electric field that depends on the coordinate along the wire as $\mathbf{E}^{loc} e^{-jq_z z}$ can be found from the boundary condition on the wire surface:

$$\alpha = \frac{I}{E_z^{loc}} = \left[\frac{\eta(k^2 - q_z^2)}{4k} H_0^{(2)} \left(\sqrt{k^2 - q_z^2}\, r_0 \right) \right]^{-1} \tag{5.166}$$

$$\approx \left[\frac{\eta(k^2 - q_z^2)}{4k} \left(1 - j\frac{2}{\pi} \left\{ \log \frac{\sqrt{k^2 - q_z^2}\, r_0}{2} + \gamma \right\} \right) \right]^{-1} \tag{5.167}$$

where $\gamma \approx 0.5772$ is the Euler constant. The approximate relation (5.167) refers to the case of thin wires with $|\sqrt{k^2 - q_z^2}\, r_0| \ll 1$. Thus, we can model the wires by lines of current with the known susceptibility (5.166).

5.5.1 Dispersion Equation

The dispersion characteristics of the media under consideration can be found as solutions of the corresponding eigenvalue problem. We will follow the deduction introduced in [30] for a three-dimensional lattice of dipoles. Assuming that an eigenwave has a plane-wave spatial dependence $e^{-j(q_x x + q_y y + q_z z)}$, we write the expression for the local electric field acting on the reference wire as the sum of the wire fields (5.164), and define the interaction constant C:

$$
\begin{aligned}
E_z^{loc} &= -\frac{\eta(k^2 - q_z^2)}{4k} \sum_{(m,n) \neq (0,0)} \left[H_0^{(2)} \left(\sqrt{k^2 - q_z^2}\, R_{m,n} \right) e^{-j(q_x am + q_y bn)} \right] I \\
&= CI
\end{aligned}
\tag{5.168}
$$

where C is the dynamic interaction constant that describes interaction effects in the infinite lattice:

$$C = -\frac{\eta(k^2 - q_z^2)}{4k} \sum_{(m,n) \neq (0,0)} \left[H_0^{(2)} \left(\sqrt{k^2 - q_z^2}\, R_{m,n} \right) e^{-j(q_x am + q_y bn)} \right] \tag{5.169}$$

Using these formulas and the definition of wire susceptibility (5.165), we can write the dispersion equation:

$$\alpha^{-1} = C \tag{5.170}$$

The main difficulty there is, of course, the calculation of the double series in the expression for the interaction constant C. Direct calculation of this series is not possible due to the fact that the series is diverging (in the absence of losses). Using summation by layers corresponding to the index m, we can rewrite the double series through terms $\beta(m)$ corresponding to each layer:

$$C(k, q_x, q_y, q_z) = \sum_{m=-\infty}^{+\infty} \beta(m) e^{-jq_x a m} \qquad (5.171)$$

$$\beta(m) = -\frac{\eta(k^2 - q_z^2)}{4k} \sum_{n=-\infty}^{+\infty} \left[H_0^{(2)} \left(\sqrt{k^2 - q_z^2}\, R_{m,n} \right) e^{-jq_y bn} \right], \quad \text{for } m \neq 0 \qquad (5.172)$$

$$\beta(0) = -\frac{\eta(k^2 - q_z^2)}{4k} \sum_{n \neq 0} \left[H_0^{(2)} \left(\sqrt{k^2 - q_z^2}\, R_{0,n} \right) e^{-jq_y bn} \right] \qquad (5.173)$$

The term $\beta(0)$ corresponding to the zero-plane can be calculated using the Poisson summation formula with singularity cancellation [31]:

$$\beta(0) = -\frac{\eta(k^2 - q_z^2)}{4k} \sum_{n \neq 0} \left[H_0^{(2)} \left(\sqrt{k^2 - q_z^2}\, b|n| \right) e^{-jq_y bn} \right] \qquad (5.174)$$

$$= -\frac{\eta(k^2 - q_z^2)}{2k} \left[\frac{1}{b\sqrt{k^2 - q_y^2 - q_z^2}} - \frac{1}{2} + \frac{j}{\pi} \left(\log \frac{b\sqrt{k^2 - q_z^2}}{4\pi} + \gamma \right) \right.$$

$$\left. + \frac{j}{b} \sum_{n \neq 0} \left(\frac{1}{\sqrt{\left(q_y + \frac{2\pi n}{b} \right)^2 + q_z^2 - k^2}} - \frac{b}{2\pi|n|} \right) \right]$$

We have already used this summation in the derivation of the averaged boundary conditions for a single wire grid; see (4.24).

If the distance between the layers is not small compared to the grid periods in each layer, it is reasonable to use the Poisson summation formula (4.11) leading to the Floquet representation for the calculation of terms $\beta(m)$ of the other layers:

$$\beta(m) = -\frac{\eta(k^2 - k_z^2)}{4k} \sum_{n=-\infty}^{+\infty} \left[H_0^{(2)} \left(\sqrt{k^2 - q_z^2}\, R_{m,n} \right) e^{-jq_y bn} \right]$$

$$= -\frac{\eta(k^2 - q_z^2)}{2kb} \sum_{n=-\infty}^{+\infty} \frac{e^{-jk_x^{(n)} a|m|}}{k_x^{(n)}} \qquad (5.175)$$

Here $k_x^{(n)}$ denotes the x-component of nth Floquet mode's wavevector:

$$k_x^{(n)} = -j\sqrt{\left(q_y + \frac{2\pi n}{b} \right)^2 + q_z^2 - k^2}, \qquad \mathrm{Re}\{\sqrt{\cdot}\} > 0 \qquad (5.176)$$

We have used the formula for the Fourier transform of the Hankel function (4.12).

Substituting (5.175) into (5.171), changing the order of summation and using the following expression for summation of the geometrical progressions [see (5.105)]:

$$\sum_{m=-\infty}^{\infty} e^{-jk_x^{(n)}a|m|-jq_z am} = \frac{j\sin k_x^{(n)}a}{\cos k_x^{(n)}a - \cos q_x a} \qquad (5.177)$$

we obtain a very useful expression for the calculation of the dynamic interaction constant:

$$C(k, q_x, q_y, q_z) = \beta(0) - \frac{\eta(k^2 - q_z^2)}{2kb} \sum_{n=-\infty}^{+\infty} \frac{1}{k_x^{(n)}} \left(\frac{j\sin k_x^{(n)}a}{\cos k_x^{(n)}a - \cos q_x a} - 1 \right) \qquad (5.178)$$

The series here converges very quickly provided the separation between layers a is not small compared with the value of b.

Substituting (5.178) together with (5.166) and (5.174) into the dispersion equation (5.170), we get

$$\frac{1}{\pi} \log \frac{b}{2\pi r_0} + \frac{1}{bk_x^{(0)}} \frac{\sin k_x^{(0)}a}{\cos k_x^{(0)}a - \cos q_x a}$$

$$+ \sum_{n\neq 0} \left(\frac{1}{bk_x^{(n)}} \frac{\sin k_x^{(n)}a}{\cos k_x^{(n)}a - \cos q_x a} - \frac{1}{2\pi|n|} \right) = 0 \qquad (5.179)$$

Note that the real part of the inverse wire susceptibility α^{-1} (5.167) cancels out with one of the terms in the formula for the interaction constant (5.174) ($-1/2$ in the brackets). The real part of α^{-1} is responsible for the field scattered by the wire, and we know that this indeed must cancel with a part of the interaction field. We have already seen similar cancellations in Chapter 4. The real parts of $\beta(0)$ and $\beta(m)$ that are left after this cancellation give plane wave fields of eigenwaves traveling in the regular lattice. The dispersion equation (5.179) is exact for thin ideally conducting wires; the only assumptions are that $r_0 \ll a, b, \lambda$, and a is not small compared with b. Let us note that this is a real-valued dispersion equation. The terms corresponding to the evanescent Floquet modes are purely real:

$$\frac{1}{bk_x^{(n)}} \frac{\sin k_x^{(n)}a}{\cos k_x^{(n)}a - \cos q_x a} = \frac{1}{b\,\mathrm{Im}\{k_x^{(n)}\}} \frac{\sinh(\mathrm{Im}\{k_x^{(n)}\}a)}{\cosh(\mathrm{Im}\{k_x^{(n)}\}a) - \cos q_x a} \qquad (5.180)$$

The obtained dispersion equation (5.179) looks asymmetric with respect to an interchange of a and b together with k_x and k_y. Obviously, it must be symmetric due to the symmetry of (5.169), from which we have started the derivation. This is not easy to prove analytically, but numerical checks show that expression (5.179) does possess this symmetry property. This means that we can use instead of (5.179) its equivalent form

$$\frac{1}{\pi}\log\frac{a}{2\pi r_0} + \frac{1}{ak_y^{(0)}}\frac{\sin k_y^{(0)}b}{\cos k_y^{(0)}b - \cos q_y b}$$

$$+ \sum_{n\neq 0}\left(\frac{1}{ak_y^{(n)}}\frac{\sin k_y^{(n)}b}{\cos k_y^{(n)}b - \cos q_y b} - \frac{1}{2\pi|n|}\right) = 0 \qquad (5.181)$$

$$k_y^{(n)} = -j\sqrt{\left(q_x + \frac{2\pi n}{a}\right)^2 + q_z^2 - k^2}, \qquad \mathrm{Re}\{\sqrt{\cdot}\} > 0 \qquad (5.182)$$

under the restriction that, in contrast to (5.179), b is not small compared with a. Thus, using one of the two equivalent forms, this restriction can be dropped.

The analyzed system can be viewed as a uniform isotropic medium (modeled by an equivalent transmission line) periodically loaded by thin grids of wires (modeled by grid impedances). This suggests that the general theory of periodically loaded transmission lines of Section 5.3 can be applied here. And indeed, the dispersion equation (5.179) can be rewritten as

$$\cos q_x a - \cos k_x^{(0)} a = j\frac{\eta}{2Z_s}\sin k_x^{(0)} a \qquad (5.183)$$

where the effective impedance for a layer of wires reads

$$Z_s = j\frac{\eta}{2}\left[\frac{bk_x^{(0)}}{\pi}\log\frac{b}{2\pi r_0} + k_x^{(0)}\sum_{n\neq 0}\left(\frac{1}{k_x^{(n)}}\frac{\sin k_x^{(n)}a}{\cos k_x^{(n)}a - \cos q_x a} - \frac{b}{2\pi|n|}\right)\right]$$
$$(5.184)$$

This impedance, without taking into account the terms corresponding to the evanescent modes, does not depend on the propagation constant:

$$Z_s \approx j\frac{\eta}{2}\frac{bk_x^{(0)}}{\pi}\log\frac{b}{2\pi r_0} \qquad (5.185)$$

It is simply the grid impedance Z_g (4.39) of an individual planar array of wires. We see again that in the plane-wave-interaction approximation (only the fundamental Floquet modes between the wire layers), the dispersion equation can be solved analytically, and we have the classical result presented in Section 5.3. But if all the Floquet harmonics are taken into account, the equivalent surface impedance (5.184) depends not only on the frequency, but also on the wavevector: $Z_s = Z_s(k, q_x, q_y, q_z)$. Interestingly, the higher-order mode influence can be included in the model through additional terms in the grid impedance of layers. After that we can still formally make use of the simple theory of periodically loaded transmission lines. This approach was probably used for the first time by J.R. Wait [53] in the theory of a wire grid near a metal plane.

Sometimes it can be useful to write a simplified dispersion equation for the axial propagation (along y-axis, $q_x = q_z = 0$) in the form:

$$\frac{1}{\pi} \log \frac{b}{2\pi r_0} = \frac{\cot(k_x^{(0)}a/2)}{bk_x^{(0)}} + \sum_{n\neq 0} \left(\frac{\cot(k_x^{(n)}a/2)}{bk_x^{(n)}} + \frac{1}{2\pi|n|} \right) \qquad (5.186)$$

$$k_x^{(n)} = -j\sqrt{\left(q_y + \frac{2\pi n}{b}\right)^2 - k^2}, \qquad \mathrm{Re}\{\sqrt{\cdot}\} > 0 \qquad (5.187)$$

or in the equivalent symmetrically transformed form:

$$\frac{1}{\pi} \log \frac{a}{2\pi r_0} + \frac{1}{ak_y^{(0)}} \frac{\sin k_y^{(0)}a}{\cos k_y^{(0)}a - \cos q_y a}$$

$$+ \sum_{n\neq 0} \left(\frac{1}{ak_y^{(n)}} \frac{\sin k_y^{(n)}b}{\cos k_y^{(n)}b - \cos q_y b} - \frac{1}{2\pi|n|} \right) = 0 \qquad (5.188)$$

$$k_y^{(n)} = -j\sqrt{\left(\frac{2\pi n}{a}\right)^2 - k^2}, \qquad \mathrm{Re}\{\sqrt{\cdot}\} > 0 \qquad (5.189)$$

In (5.186), the dependence on q_y is only through parameters $k_x^{(n)}$, but in (5.186), the values $k_y^{(n)}$ are independent on q_y. Thus, we have different in form but equivalent expressions (5.186) and (5.188) of the dispersion equation for the axial propagation. We can use any of these formulas with the same result.

Typical dispersion curves for a square grid of cylinders (in this example the filling ratio $f = \pi r_0^2/a^2 = 0.001$) are shown in Figure 5.18, where $\Gamma = (0,0,0)$, $X = (\pi/a, 0, 0)$, and $M = (\pi/a, \pi/a, 0)$ are points in the first Brillouin zone. Here, together with the thick lines representing the dispersion curves for the wire medium, the dispersion curves for free space are presented for comparison (as thin lines).

5.5.2   Reflection Coefficient

Let us consider a plane interface between a half-space filled with a wire medium ($x \geq 0$ or index $m \geq 0$) and free space. Suppose that an incident plane electromagnetic wave $\mathbf{E}e^{-j(k_x x + k_y y + k_z z)}$ is exciting the medium. To solve for the reflection, it is convenient to split the incident electric field vector into the longitudinal and transverse parts with respect to the wire

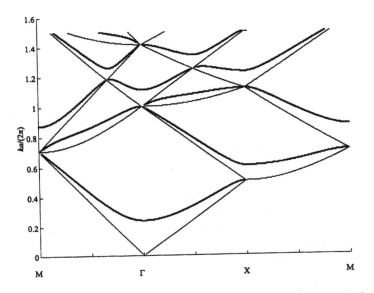

Figure 5.18 Dispersion curves for a square grid of cylinders with the filling ratio $f = 0.001$ (thick lines) and dispersion curves for free space (thin lines).

axis. Obviously, under our assumption of thin wires, the wave whose electric field is orthogonal to the wires does not see the grid, and the following results concern the longitudinal part only. If we neglect the transition layer (see Section 5.2.3) and assume that even the first row of wires is excited in the same way as the wires in the bulk, we can use (5.116) to find the reflection coefficient:

$$R = \frac{\sin[(k_x - q_x)a/2]}{\sin[(k_x + q_x)a/2]} \tag{5.190}$$

This value is real for propagating modes (q_x is purely real). For decaying modes inside a stopband (when q_x is purely imaginary), we have full power reflection: $|R| = 1$.

As we have already explained, the correct sign of q_x should be chosen in (5.190). We solve the dispersion equation for real propagation constants $\tilde{q}_x$ in the range $[0 \ldots \pi/a]$, and for the regime with a single propagating mode $ka/(2\pi) < 1$ we should take $q_x = \tilde{q}_x$, if $ka/(2\pi) < 0.5$, but $q_x = -\tilde{q}_x$, if $ka/(2\pi) > 0.5$. From the two solutions having a nonzero imaginary part of the propagation constant, we should choose decaying ones.

Influence of the transition layer, neglected in deriving (5.190), can be included in the analysis considering not only propagating but also evanescent modes. The evanescent modes do not influence the absolute value of the reflection coefficient but influence its phase. The generalized formula for the reflection coefficient was established in [30]:

$$R = e^{-jk_x a} \prod_{n=1}^{+\infty} e^{jk_x a} \frac{\sin[(k_x - q_x^{(n)})a/2]}{\sin[(k_x + q_x^{(n)})a/2]} \tag{5.191}$$

where $q_x^{(n)}$ are the solutions of the dispersion equation (5.179) with $q_y = k_y$, $q_z = k_z$. The product in (5.191) includes all the modes propagating *into* the half-space filled by the lattice. It means that we should take the correct sign of $q_x^{(n)}$ (corresponding to the direction of the field decay from the source into the half-space), as was explained in connection with the simplified equation (5.190).

Equations (5.190) and (5.191) are valid only if magnetic polarization in the medium can be neglected. This is obvious from the fact that (5.190) for dense grids reduces to (5.121), which is valid only if $\mu_{\text{eff}} = \mu_0$. This is clearly a correct assumption for the wire medium made up of straight wires, but the restriction is important for various generalizations of the theory.

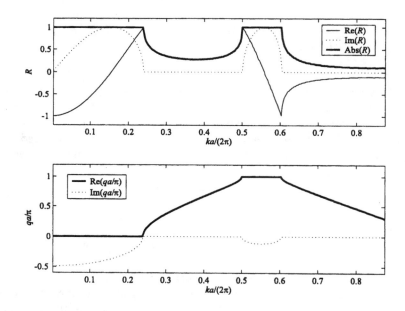

Figure 5.19 Dependence of the reflection coefficient from a half-space filled by a square grid of cylinders with the filling ratio $f = 0.001$, and the corresponding propagation constant on the normalized frequency.

The reflection coefficient from a half-space filled by the same wire medium as in Figure 5.18 (a square grid of cylinders with the filling ratio $f = 0.001$) is shown in Figure 5.19 as a function of the normalized frequency

$ka/(2\pi)$. Note the similarity with transmission lines loaded by inductive loads. The reflection coefficient equals -1 at very low frequencies. Within the low-frequency stopband it changes its phase, so that at the top edge of the bandgap at $ka/(2\pi) \approx 0.25$, the reflection coefficient becomes equal to $+1$. In other words, the interface behaves as a metal wall at $ka \to 0$, but it reflects as a magnetic wall at frequencies close to the first propagation window. The reflection coefficient remains positive in the first propagation region $0.25 < ka/(2\pi) < 0.5$, and it again becomes equal to $+1$ at the bottom edge $ka = \pi$ of the second stopband, and after that it changes phase inside the stopband at frequencies $ka = \pi$ to $ka/(2\pi) \approx 0.6$, and becomes equal to -1 at the top edge. In the second passband window from $ka/(2\pi) \approx 0.6$ to $ka = 2\pi$, the reflection coefficient is negative.

5.5.3   Dense Arrays

If both periods of the array a and b are very small compared to the wavelength in the structure, we might expect that a certain homogenization of this "material" is possible. From the geometry, we would expect that the wire medium could be modeled by a uniaxial permittivity dyadic, because for dense grids there is only one preferred direction: along the wires. Let us proceed along the same way as in Section 5.4.1: First, we make an assumption that the effective permittivity dyadic is of the form

$$\overline{\overline{\epsilon}}_{\text{eff}} = \epsilon_t \overline{\overline{I}}_t + \epsilon_z \mathbf{z}_0 \mathbf{z}_0 \tag{5.192}$$

We know that the dispersion relations on such media read (5.136)–(5.137)

$$q_z^2 + q_x^2 + q_y^2 = \omega^2 \mu_0 \epsilon_t, \qquad \text{TE modes} \tag{5.193}$$

$$q_z^2 + \frac{\epsilon_t}{\epsilon_z}(q_x^2 + q_y^2) = \omega^2 \mu_0 \epsilon_t, \qquad \text{TM modes} \tag{5.194}$$

For the wire media we are interested only in TM modes, since the TE modes have no electric field component along $\mathbf{z}_0$ and do not feel the wires at all. Next, we simplify the dispersion relation for wire media and try to reduce it to the form of (5.194), in order to identify the effective permittivities.

If the grid is dense compared with the wavelength, the dispersion equation (5.179) can be simplified using the Taylor expansion of sine and cosine functions of small arguments and analytically solved. The result is the following:

$$q^2 = q_x^2 + q_y^2 + q_z^2 = k^2 - k_p^2 \tag{5.195}$$

where

$$k_p^2 = \frac{2\pi/(ab)}{\log \dfrac{b}{2\pi r_0} + \displaystyle\sum_{n=1}^{+\infty} \left(\dfrac{\coth \frac{\pi n a}{b} - 1}{n} \right) + \dfrac{\pi a}{6b}} \tag{5.196}$$

depends only on the grid geometry. Before considering the homogenization problem further, let us discuss the last result. From (5.195) we see that there is a stopband at low frequencies with the upper boundary at the frequency corresponding to k_p: For $k < k_p$, the propagation factor $q = -j\sqrt{k_p^2 - k^2}$ is an imaginary number and the wave exponentially decays. For $k > k_p$, we have a propagating wave with $q = \sqrt{k^2 - k_p^2}$. It is interesting to observe that k_p as a function of the lattice constants a and b is symmetric: $k_p(a,b) = k_p(b,a)$. This follows from the symmetry property of the dispersion equation (5.179) described above, and also can be easily numerically checked. This fact is physically sound and obvious, but it is not so easy to see it from expression (5.196). For the square grid ($a = b$), relation (5.196) simplifies, and we have

$$k_p^2 = \frac{2\pi/a^2}{\log \dfrac{a}{2\pi r_0} + 0.5275} \qquad (5.197)$$

Let us now return to the problem of effective parameters at low frequencies. First, we note that since we neglect the wire polarization in the transverse directions, the parameter ϵ_t must be equal to ϵ_0. Thus, (5.194) for the model uniaxial material becomes

$$\epsilon_z q_z^2 + q_x^2 + q_y^2 = \omega^2 \mu_0 \epsilon_z \qquad (5.198)$$

Comparing with the low-frequency dispersion (5.195) of the wire medium, we see that it is not compatible with the dispersion equation for homogeneous uniaxial media [54]. Indeed, there does not exist an effective parameter $\epsilon_z(\omega)$, such that (5.195) would take the same form as (5.198). However, if we allow the effective permittivity to depend on the propagation factor q_z and choose

$$\epsilon_z(\omega, q_z) = \epsilon_0 \left(1 - \frac{k_p^2}{k^2 - q_z^2} \right) \qquad (5.199)$$

the actual and model dispersion relations coincide. The conclusion is that the considered wire media can be described by the uniaxial permittivity dyadic (5.192) and considered as an effectively homogeneous material at low frequencies, but the axial permittivity ϵ_z must be a nonlocal parameter of the form (5.199). The material is spatially dispersive, and it is a very peculiar case when the spatial dispersion effects remain strong even at extremely low frequencies [54].

It is possible to transit (5.199) from the spectral domain $(\mathbf{q}, \omega)$ to the physical domain $(\mathbf{r}, t)$. The following nonlocal material equation can be derived from (5.199) using the double Fourier transform:

$$\mathbf{D}(x, y, z) = \epsilon_0 \mathbf{E}(x, y, z) + \frac{\epsilon_0 k_p^2 c}{4} \mathbf{z}_0 \int\limits_{-\infty}^{t} \int\limits_{z-c(t-t')}^{z+c(t-t')} E_z(x, y, z', t') \, dz' dt' \qquad (5.200)$$

where $c = 1/\sqrt{\epsilon_0 \mu_0}$ is the speed of light in the host matrix that is assumed to be dispersionless. Here the area of integration in the (z, t) plane is the light cone $|z - z'| < c(t - t')$. In other words, the kernel in the Fourier convolution is $u[c(t - t') - |z - z'|]$, where $u(x)$ is the Heaviside step function. It means that the point (x, y, z) inside the wire medium (described as a dispersive continuum) at moment t is affected by the z-components of electric fields coming from the domain $[x, y, z \pm c(t - t')]$ surrounding (along the wire axis) this point during all the past time $(t' < t)$. The contributions of these partial fields are equal to each other, and the contributions of the near domains of the wire do not dominate.

Spatial dispersion in inhomogeneous structures usually appears in situations when the wavelength in the host medium is comparable with the characteristic dimensions of the structure. In wire media of infinite (practically, very long as compared to the lattice period) wires, spatial dispersion appears to remain a very strong factor even at very low frequencies.

The effective permittivity (5.199) is a negative number at low frequencies, when $k_p^2 > k^2 - q_z^2$, and in this sense the wire medium is similar to the free-electron plasma (this is the reason for the notation k_p, as the corresponding frequency plays the role of the equivalent *plasma frequency*). The wire medium is a component of the known experimental realization of artificial media with negative real parts of the material parameters [55]. However, let us stress that this analogy between plasma and wire media is rather limited because of strong spatial dispersion effects in wire media that are absent in free-electron plasmas.

5.5.4 Quasistatic Modeling of Wire Media

The full-wave model of wire media developed above is limited to the case of thin ideally conducting wires. Extensions of the model to more complicated structures are possible, but they can lead to difficulties in obtaining analytical and easily interpretable solutions.[12] On the other hand, if we restrict the analysis to low frequencies and are interested in the propagation only in the transverse plane $(q_z = 0)$, it is possible to introduce a simple and very useful model [56] that can also cover thick, lossy, or loaded wires. In many applications this model is enough to understand the properties of artificial materials that contain arrays of long wires.

The Effective Permittivity

The geometry of the problem is again illustrated in Figure 5.20. Consider a layer of wire medium located between two imaginary planes orthogonal to the wires. The planes are positioned at $z = 0$ and $z = d$. Let us assume that the external electric field is parallel to the wires (axis z). If distance d

---

[12] An extension to thin loaded wires is possible, and we will consider that in Chapter 6.

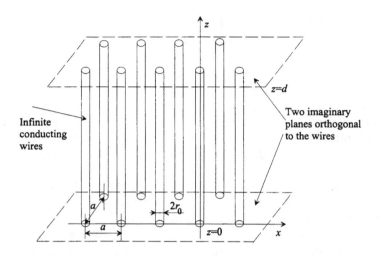

Figure 5.20 Geometry of the problem: an infinite periodical array of conducting wires.

is much smaller than the wavelength λ, the space between the two cutting planes can be considered as a plane capacitor, and the permittivity of the effective medium formed by conducting wires inside the layer can be found from simple quasistatic considerations.

Let us assume that the voltage between the planes at $z = 0$ and $z = d$ equals U. Electric current I flows in each wire of the medium, which reads, from the circuit theory,

$$U = Ij\omega Ld \qquad (5.201)$$

where L is the total (including mutual coupling between wires) inductance of one wire per unit length. On the other hand, the voltage is connected with the average electric field in the medium (its z-component):

$$U = E_z d \qquad (5.202)$$

From the above relations, we have

$$E_z = j\omega L\, I \qquad (5.203)$$

The electric displacement vector $\mathbf{D}$ is related to the electric field and the polarization inside a medium as follows:

$$\mathbf{D} = \epsilon_0 \mathbf{E} + \mathbf{P} \qquad (5.204)$$

For our case

$$\mathbf{P} = \frac{\mathbf{J}}{j\omega} = \mathbf{z}_0 \frac{I}{j\omega a^2} = -\mathbf{z}_0 \frac{E_z}{\omega^2 a^2 L} \qquad (5.205)$$

Finally,

$$\mathbf{D} = \left(\epsilon_0 \overline{\overline{I}} - \frac{\mathbf{z}_0 \mathbf{z}_0}{\omega^2 a^2 L} \right) \cdot \mathbf{E} \tag{5.206}$$

Let us now find the wire inductance per unit length L. To do that we need to calculate magnetic flux per a unit length segment of a wire, so we need to estimate the magnetic field distribution in the infinite array of wires. At this stage, we note that due to the symmetry of the problem, the magnetic field is zero at the middle points between wires. On the other hand, if the wires are thin, the main contribution into the flux comes from the area close to the wires. These observations bring us to the following estimation for the magnetic field in the space between two neighboring wires located at $x = 0$, $y = 0$ and $x = a$, $y = 0$:

$$H_y = \frac{I}{2\pi} \left(\frac{1}{x} - \frac{1}{a - x} \right) \tag{5.207}$$

This assumes the quasistatic formula for the magnetic field of the wires. The first term is simply the quasistatic field of the wire positioned at $x = 0$, $y = 0$, and the second term is the field of the neighboring wire, which we include to make the total field zero at the symmetry lines. Now we can calculate the magnetic flux per unit length:

$$\Psi = \mu_0 \int_{r_0}^{a/2} H_y(x)\, dx = \frac{\mu_0 I}{2\pi} \log \frac{a^2}{4 r_0 (a - r_0)} \tag{5.208}$$

where r_0 is the wire radius. Hence, for the inductance we obtain

$$L = \frac{\mu_0}{2\pi} \log \frac{a^2}{4 r_0 (a - r_0)} \tag{5.209}$$

and for the permittivity

$$\overline{\overline{\epsilon}} = \epsilon_0 \left[\overline{\overline{I}} - \frac{2\pi \mathbf{z}_0 \mathbf{z}_0}{(ka)^2 \log \dfrac{a^2}{4 r_0 (a - r_0)}} \right] = \epsilon_0 \left(\overline{\overline{I}} - \mathbf{z}_0 \mathbf{z}_0 \frac{k_p^2}{k^2} \right) \tag{5.210}$$

where $k = \omega \sqrt{\epsilon_0 \mu_0}$. The cutoff wavelength $\lambda_p = 2\pi / k_p$ is obviously

$$\lambda_p = a \sqrt{2\pi \log \frac{a^2}{4 r_0 (a - r_0)}} \tag{5.211}$$

The dependence of the normalized cutoff wavelength on the normalized wire radius is plotted in Figure 5.21. If r_0 tends to zero, the permittivity tends to ϵ_0, as it should be. The other limit is when r_0 tends to $a/2$. In this case spacing between wires disappears, and the $\mathbf{z}_0 \mathbf{z}_0$ component of the permittivity becomes infinite (simply ideally conducting material between the two planes).

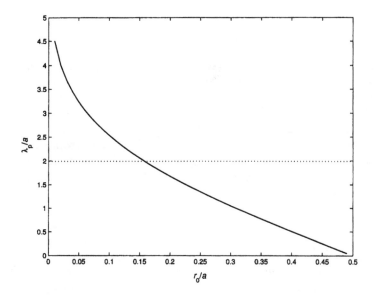

Figure 5.21 Dependence of the cutoff wavelength on the normalized wire radius. The quasistatic assumptions are not applicable below the dotted line $\lambda = 2a$.

Limitations of the Model and Comparison with Other Models

The validity region of the quasistatic approximation for the magnetic field surrounding a wire can be estimated as $a/2 < \lambda/4$ (i.e., $\lambda > 2a$), where λ is the free-space wavelength. This limiting border is shown in Figure 5.21 by a horizontal line. The other limitation occurs due to spatial dispersion effects. For large absolute values of the permittivity, the phase and the amplitude of the wave field inside the medium vary quickly. In our case it leads to a strong nonuniformity in the current distribution among the wires. We may think that it can drastically change the local field and the magnetic flux of a wire, resulting in a strong dependence of the permittivity on the wavenumber. But let us consider this more thoroughly.

Suppose that a wave is propagating (or decaying, if $\epsilon < 0$) along the y direction. Then, the phases (amplitudes) of the wire currents change along the direction of propagation (or decay). However, we see that it is again possible to calculate the flux in the space between two neighboring wires according to (5.207), because there are no variations of phases (amplitudes) along the x-axis and our quasistatic assumptions are still valid here. Of course, a similar conclusion holds for orthogonally propagating or decaying waves, providing the integration (5.208) is made along the y-axis. From this we conclude that dependence of the permittivity on the wave vector is probably important only for very large absolute values of the permittivity and high frequencies, where the present quasistatic approach is not valid anyway.

Let us compare the quasistatic result with the low-frequency asymptotic of the full-wave theory for thin wires (5.199) with the cutoff wavenumber k_p given by (5.197). First, we note that the quasistatic model cannot take into account spatial dispersion. That is why the dependence on the wavenumber q_z is missing in (5.210). Thus, in this case the quasistatic results can be used only for waves with $q_z = 0$, that is, propagating in the transverse plane.[13]

Substituting k_p, (5.199) (with $q_z = 0$) can be written as

$$\bar{\bar{\epsilon}} = \epsilon_0 \left[\bar{\bar{I}} - \frac{2\pi \mathbf{z}_0 \mathbf{z}_0}{(ka)^2 \left(\log \frac{a}{2\pi r_0} + 0.5275 \right)} \right] \qquad (5.212)$$

We observe that the only difference between (5.210) and (5.212) is in the slow-varying logarithmic term. To make a comparison, we plot the effective permittivity (Figure 5.22) calculated using the quasistatic method (5.210) (solid curves), full-wave model for thin wires (5.212) (dotted lines), and the equivalent transmission-line model [39, 40] (dashed lines). The last model considers planar layers of wires as insertions in an equivalent transmission line modeling free space between the grid planes, and the effective permittivity is derived by approximating the dispersion equation for periodically loaded transmission lines (5.106). The result of that approach is the same as (5.212), but the correction term 0.5275 that describes the influence of the higher-order Floquet modes between the planes of wires is missing.

All three models agree very well for very thin wires, because in this case the interaction field in the array is very small compared to the local field of a single thin wire. For larger wire radii we observe that the full-wave analysis for thin wires as well as the transmission-line model lead to considerable errors. Those models assume that the wires are infinitely thin current lines, so the validity region of the numerical model is $r_0/a \ll 1$. The quasistatic model correctly takes into account the wire diameter in the quasistatic frequency region. The transmission-line model [39] is more suitable for the case when one of the two array periods is much larger than the other one.

5.6 THREE-DIMENSIONAL ELECTROMAGNETIC CRYSTALS

Consider a regular array ("crystal") formed by small dipole particles (Figure 5.23). Here we will design an analytical model with the goal of demon-

---

[13]Remember that here we analyze only TM waves, which interact with the wires. TE waves propagate as in free space, and there is no such limitation on the medium model for TE waves.

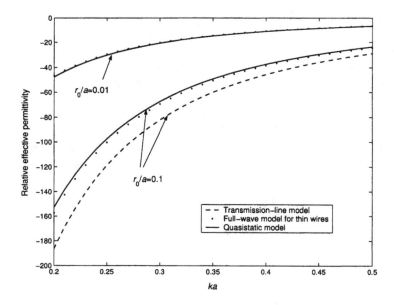

Figure 5.22 Comparison between the analytical and analytical-numerical models.

strating the main physical features of electromagnetic crystals. Our intent is to present a simple and physically clear model that shows how the periodicity of the material affects the electromagnetic properties in wide frequency ranges. To allow an analytical solution, we consider only small particles modeled by dipoles. In the general case, the problem has to be attacked numerically.

Analytical models of crystals and other periodic structures are usually considered under one of the following assumptions: Either the periods are very small compared to the wavelength (effective medium case, see Section 5.2), or the field structure in one period of the array can be represented by two propagating waves (e.g., [57,58]). The last assumption is valid when one of the periods is large compared to the other two, and the system can be modeled by an equivalent transmission line with periodical insertions (see Section 5.3). The last method is widely used in the analysis and synthesis of microwave filters and frequency-selective surfaces. In the general case of arbitrary shapes of inclusions, computationally demanding numerical methods are used.

In this chapter we present an analytical model [59] that can be used from very low frequencies up to the resonance and appearance of stopbands (called *photonic bandgaps* in the optical literature). The main limitation is the assumption about the small size of individual inclusions.

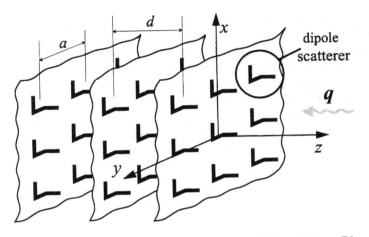

Figure 5.23 Geometry of an infinite array of small particles. Plane wave propagates along axis z.

5.6.1 Basic Assumptions and Main Relations

Consider a regular and infinite three-dimensional array of particles that can be modeled by electric or magnetic dipoles (Figure 5.23). The system of particles is periodic along all three Cartesian axes. The periods along x and y are equal to a; the period along z is d. In this section we consider only the axial propagation of waves (along axis z) and neglect absorption in the inclusions. If $d \gg a$, each "layer" of dipoles in the plane $x - y$ can be replaced by an equivalent grid impedance that we developed in Chapter 4, Section 4.5. The boundary condition (4.95) for a planar array of particles is a relation between the averaged[14] over the array plane electric field $\mathbf{E}$ and the averaged surface current density $\mathbf{J} = j\omega \mathbf{p}/a^2$ ($\mathbf{p}$ is the particle dipole moment):

$$\mathbf{E} = Z_g \mathbf{J}, \quad Z_g = \frac{a^2}{j\omega}\left[\frac{1}{\alpha} - j\frac{k^3}{6\pi\epsilon_0} - \operatorname{Re}\beta\right] \qquad (5.213)$$

Here α is the single particle polarizability, and β is the interaction constant for planar arrays of dipoles, which defines the contribution to the local field $\mathbf{E}_{\text{loc}}$ from the dipoles located in the same plane:

$$\mathbf{p} = \alpha\mathbf{E}_{\text{loc}}, \qquad \mathbf{E}_{\text{loc}} = \mathbf{E}_{\text{ext}} + \beta\mathbf{p} \qquad (5.214)$$

where $\mathbf{E}_{\text{ext}}$ is the external field. Since the wave propagates along axis z, all dipole moments in every $(x - y)$-plane are equal. Approximate analytical expressions for the interaction constant have already been established in

---

[14]As is usually done in the effective medium theories, we denote here the averaged electric field $\widehat{\mathbf{E}}$ simply as $\mathbf{E}$, as well as the averaged current density $\widehat{\mathbf{J}} = \mathbf{J}$, and so forth.

Chapter 4. The impedance model can be applied when the array of dipoles is separated from other inhomogeneities (in our case, other planes of dipoles) by a distance larger than the distance between inclusions in the plane. In that case, the evanescent modes decay and only the averaged polarization that creates a propagating wave determines the field of the array. Thus, if $d \gg a$, we can use the theory of periodically loaded transmission lines (Section 5.3) and write the eigenvalue equation for three-dimensional arrays in the same form as for one-dimensional transmission lines:

$$\cos qd = \cos kd + j\frac{\eta}{2Z_g} \sin kd \qquad (5.215)$$

Here q is the propagation factor, and $\eta = \sqrt{\mu_0/\epsilon_0}$ and $k = \omega\sqrt{\epsilon_0\mu_0}$ are the wave impedance and the wavenumber in the matrix medium in which the particles are located.

If the assumption $d \gg a$ is not valid, the field between the layers has to be expanded into the whole Floquet series (e.g., [31, Section 9.7]), which complicates the analysis. On the other hand, if the spacings between inclusions are very small in wavelengths, various mixing rules can be applied to determine the effective permittivity (and the propagation constant) [15]. We will show how to extend the simple transmission-line approach to account for higher-order Floquet harmonics in three-dimensional arrays.

5.6.2   General Eigenvalue Equation

We start from deriving the eigenvalue equation (5.215) by another method, which will make clear how to include the higher-order modes in the model. This derivation will be similar to that in Section 5.3, as it is based on the summation of fields created by all the layers of particles. Looking for eigensolutions in the absence of external fields, we can write for an arbitrarily chosen dipole in the array (position $z = 0$)

$$\mathbf{p}(0) = \alpha\mathbf{E}_{\text{loc}} = \alpha \sum_{n=-\infty}^{\infty} \beta(nd)\mathbf{p}(nd) \qquad (5.216)$$

In infinite regular arrays, because of the Floquet theorem (4.2),

$$\mathbf{p}(nd) = \mathbf{p}(0)e^{-jnqd} \qquad (5.217)$$

where q is the unknown propagation factor we want to find.

$\beta(nd)\mathbf{p}(nd)$ gives the field generated by an array of dipoles $\mathbf{p}$ at distance nd from the array plane. Let us first assume that the distance d between the planes is much larger than the grid period a. In this case the field generated by the array is practically a plane wave; thus,

$$\beta(nd) = -j\frac{\eta\omega}{2a^2}\exp(-jk|n|d), \qquad n \neq 0 \qquad (5.218)$$

For $n = 0$ we will use (4.90) that includes local reactive fields as well. Under this assumption, the local field is

$$\mathbf{E}_{\text{loc}} = \left[\frac{\omega}{a^2} \frac{\eta}{4} \left(\frac{\cos kR_0}{kR_0} - \sin kR_0 \right) + j \frac{k^3}{6\pi\epsilon_0} \right] \mathbf{p}(0) \qquad (5.219)$$

$$- j \sum_{n=-\infty}^{\infty} \frac{\eta}{2} \frac{\omega}{a^2} \exp(-jk|n|d) \exp(-jqnd)\mathbf{p}(0) \qquad (5.220)$$

We have already done this summation (5.105):

$$\sum_{n=-\infty}^{\infty} \exp(-jk|n|d) \exp(-jqnd) = \frac{j \sin kd}{\cos kd - \cos qd} \qquad (5.221)$$

Because

$$\mathbf{p}(0) = \alpha \mathbf{E}_{\text{loc}} \qquad (5.222)$$

we can now write the eigenvalue equation:

$$\frac{1}{\alpha} = \frac{\omega}{a^2} \frac{\eta}{4} \left(\frac{\cos kR_0}{kR_0} - \sin kR_0 \right) + j \frac{k^3}{6\pi\epsilon_0} + \frac{\eta}{2} \frac{\omega}{a^2} \frac{\sin kd}{(\cos kd - \cos qd)} \qquad (5.223)$$

Let us study arrays of lossless inclusions. There the imaginary part of $1/\alpha$ is given by (4.82), and we see that in the eigenvalue equation (5.223) the imaginary parts cancel out. Thus, making use of (4.90), we can rewrite the equation as

$$\text{Re} \left\{ \frac{1}{\alpha} - \beta(0) \right\} = \frac{\eta}{2} \frac{\omega}{a^2} \frac{\sin kd}{(\cos kd - \cos qd)} \qquad (5.224)$$

Multiplying this equation by $a^2/(j\omega)$, we can express the relation in terms of the grid impedance of dipole layers (5.213):

$$Z_g = \frac{a^2}{j\omega} \sum_{n=-\infty}^{\infty} \beta(nd) = -\frac{\eta}{2} \frac{j \sin kd}{(\cos kd - \cos qd)} \qquad (5.225)$$

where for $n \neq 0$ the interaction constants $\beta(nd)$ are replaced by their far-zone values (5.218). The last relation is equivalent to the eigenvalue equation (5.215) that we already know from the theory of loaded transmission lines:

$$\cos qd = \cos kd + j \frac{\eta}{2Z_g} \sin kd \qquad (5.226)$$

In the lossless case, this is a real-valued equation since Z_g is an imaginary number. If dissipative losses are present in the particles, the equation becomes complex-valued since the dissipation loss term remains in $1/\alpha$ in the left-hand side of (5.223). The same conclusion is true for lossy matrices, since there the value of the sum (5.221) is a complex number.

Now it becomes clear how we can generalize the result, taking into account all near-field terms (in other words, all evanescent modes). We have already taken into account the propagating part of the field; now we should add the remaining parts of the interaction constants $\beta(nd)$ in (5.225). Thus, in the general case, we get

$$Z_g = -\frac{\eta}{2}\frac{j\sin kd}{(\cos kd - \cos qd)} \tag{5.227}$$

$$+\frac{a^2}{j\omega}\sum_{n=-\infty, n\neq 0}^{\infty}\left[\mathrm{Re}\left\{\beta(nd)\right\} + \frac{\eta}{2}\frac{\omega}{a^2}\sin k|n|d\right]\exp[-jqnd]$$

The term with $\sin k|n|d$ compensates the real part of the plane-wave contribution already taken into account in the first sum. A similar result (in its physical meaning) was established in [30, Eq. (2.7)]. In that paper a quasistatic model was built, and the local field was considered as the sum of the averaged field and the plane-wise sum of the static dipole fields. The averaged field was found as a solution to the wave equation that gives the same result as the summation of plane-wave fields generated by the averaged polarizations of the planes (5.221).

From (5.227), we conclude that the eigenvalue equation taking into account the evanescent modes can be written in the familiar form (5.215) with the sheet impedance (5.213) replaced by

$$Z_g \to Z_g^{\mathrm{eq}} = Z_g + j\frac{a^2}{\omega}\sum_{n=1}^{\infty}\left[2\,\mathrm{Re}\left\{\beta(nd)\right\} + \frac{\eta\omega}{a^2}\sin k|n|d\right]\cos qnd \tag{5.228}$$

(we have made use of the fact that $\mathrm{Re}\left\{\beta(nd)\right\}$ is an even function of n). In a sense we can say that the equivalent sheet impedance Z_g^{eq} takes into account higher-order mode interactions between layers of dipoles in infinite periodic arrays. Finally, the eigenvalue equation reads

$$\cos qd = \cos kd + j\frac{\eta}{2Z_g^{\mathrm{eq}}}\sin kd \tag{5.229}$$

or, substituting Z_g^{eq},

$$\cos qd = \cos kd - \frac{\sin kd}{\frac{2a^2}{\eta\omega}\left[\frac{1}{\alpha} - j\frac{k^3}{6\pi\epsilon_0} - \mathrm{Re}\{\beta(0)\}\right] - S} \tag{5.230}$$

where

$$S = \sum_{n=1}^{\infty}\left[\frac{4a^2}{\eta\omega}\,\mathrm{Re}\left\{\beta(nd)\right\} + 2\sin knd\right]\cos qnd \tag{5.231}$$

Series S in the denominator converges very rapidly, especially for $ka \ll 1$, because for large nd

$$\frac{4a^2}{\eta\omega}\,\mathrm{Re}\left\{\beta(nd)\right\} \to -2\sin knd + \frac{(ka)^2}{\pi knd}\cos knd + O(knd)^{-2} \tag{5.232}$$

Moreover, in accord with our expectations, the series term gives in fact a small correction to the denominator, so the equation can be easily solved by the recursive method.

5.6.3 Frequency Dependence of the Effective Refraction Index

Let us now study the frequency behavior of the effective response of dipole arrays. The method developed above allows us to take into account the dynamic interaction field, and to complete the analysis we need a dynamic (full-wave) expression for the inclusion polarizability. In the following examples we will consider arrays of small dielectric spheres. Because the inclusion size is much smaller than the grid periods (otherwise the dipole approximation is not justified), we use here an approximate relation for the dielectric sphere polarizability, available in the literature [60, 61]. Under the assumption that the sphere radius r is much smaller compared to the wavelength in the surrounding medium ($kr \ll 1$), the following expression for the polarizability is given in [61]:

$$\alpha = 3V\epsilon_0 \frac{\epsilon - \epsilon_0}{\epsilon + 2\epsilon_0} \left(1 - 3\frac{\epsilon - \epsilon_0}{\epsilon + 2\epsilon_0} \left[G_1(kr) + \frac{\epsilon}{\epsilon_0} G_2(kr) \right] \right)^{-1} \qquad (5.233)$$

where

$$G_1(kr) = \frac{2}{3} \left[(1 + jkr)e^{-jkr} - 1 \right] \qquad (5.234)$$

$$G_2(kr) = \left[1 + jkr - \frac{7}{15}(kr)^2 - j\frac{2}{15}(kr)^3 \right] e^{-jkr} - 1 \qquad (5.235)$$

and $k = \omega\sqrt{\epsilon_0\mu_0}$.

The first example [59] illustrates different models in estimating the field interactions between inclusions in three-dimensional arrays. First we note that the dynamic correction to the single sphere polarizability (5.233) already gives a reasonable approximation, although the particle interactions are treated in the quasistatic manner as in the standard Maxwell Garnett approach. We also use the same inclusion polarizability (5.233) in calculations of all the other curves presented in Figure 5.24. If we leave only the fundamental propagating mode in the model of interactions between the planes, that is, skip the series in (5.230), there is an error in the static limit, because even the static field of arrays at distances nd from the reference particle have been neglected (dashed line in Figure 5.24). It is interesting to note how the effects of higher-order modes interactions change with the frequency. This can be seen from comparing the solid line (full-wave solution) and the dotted line. In calculations of the last curve, only static terms in the series in (5.230) were retained. We observe that the results fall in between the results of the full-wave model (solid curve) and the dynamically corrected Maxwell

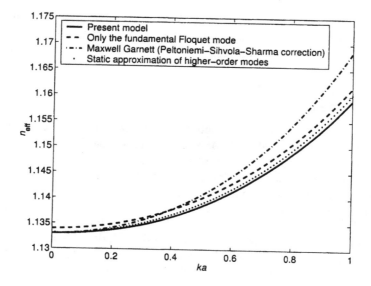

Figure 5.24 Effective refractive index q/k for a three-dimensional array of dielectric spheres with cubic cells. Permittivity of spheres is $\epsilon = 20\epsilon_0$, volume fraction of inclusions $f = 0.1$, cubic cells with the side d.

Garnett approximation (dash-dotted line). The solution of the quasistatic equation [59, Eq. (27)] gives the result that is close to the Maxwell Garnett approximation with the dynamically corrected polarizability of the spheres.

Wide frequency band behavior is illustrated in Figure 5.25. Here we have assumed that $d = 10a$. The full-wave results have been calculated as solutions to the eigenvalue equation with the dynamically corrected sphere polarizability (5.233). As should be expected, in this case of large distances between planes ($d = 10a$), evanescent modes give a very small correction at high frequencies, in contrast to the previous example with cubic cells.

PROBLEMS

5.1 Derive (5.35).

5.2 To persuade yourself that the excitation by quasiuniform magnetic fields is indeed correctly described as the excitation by the associated electric field, consider a simple example of a rectangular loop. Find the current induced by quasistatic magnetic fields by the method in Section 5.1.3. Assume that the loop is in the field of a standing electromagnetic wave and calculate the induced electromotive force without a recourse to the Stokes theorem. Prove that in this example the electromotive force does not depend on the z-axis coordinate provided that

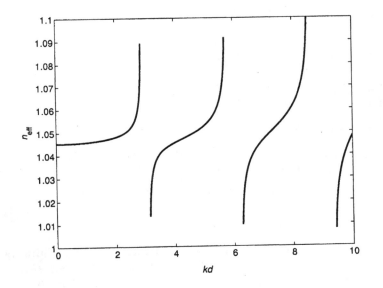

Figure 5.25 Effective refractive index for a periodic arrangement of lattices of dielectric spheres. Sphere permittivity $\epsilon = 20\epsilon_0$, sphere radius $r = 0.4a$, the distance between arrays $d = 10a$.

the axis belongs to the loop plane, that is, does not depend on the direction, in which the standing wave "propagates."

5.3 Repeat the previous problem for wire particles in the form of letters H and Γ. Discuss difficulties in introducing particle polarizabilities.

5.4 Derive (5.106) using the method in Section 5.4 (introducing an equivalent transmission matrix for a single period).

5.5 Derive (5.109).

5.6 Complement (5.113) with the corresponding relation for the distribution of current. Formulate the boundary conditions for the voltage and current on the two sides of one period (Figure 5.11) and derive the dispersion relation (5.106) from these conditions.

5.7 Study how the frequency boundaries of stopbands in periodical capacitively loaded transmission lines depend on the loads.

5.8 Study the dispersion properties of one-dimensional electromagnetic crystals (Section 5.4) for oblique propagation of waves. Determine how the widths of the stopbands depend on the propagation angle for the two eigenpolarizations.

5.9 Derive (5.200).

References

[1] Pearsall, J. (ed.), *The New Oxford Dictionary of English,* Oxford, England: Oxford University Press, 1998.

[2] http://plastics.about.com/library/glossary/blpreface.htm

[3] Jaggard, D.L., A.R. Mickelson, and C.T. Papas, "On Electromagnetic Waves in Chiral Media," *Applied Physics,* Vol. 18, 1979, pp. 211-216.

[4] Saadoum, M.M.I., and N. Engheta, "A Reciprocal Phase Shifter Using Novel Pseudochiral or Omega Medium," *Microwave and Optical Technology Lett.,* Vol. 5, 1992, pp. 184-188.

[5] Tretyakov, S.A., et al., "Analytical Antenna Model for Chiral Scatterers: Comparison with Numerical and Experimental Data," *IEEE Trans. Antennas and Propagation,* Vol. 44, No. 7, 1996, pp. 1006-1014.

[6] Schelkunoff, S.A., and H.T. Friis, *Antennas: Theory and Practice,* New York: John Wiley & Sons, 1952.

[7] King, R.W.P., and C.W. Harrison, *Antennas and Waves: A Modern Approach,* Cambridge, MA, and London, England: The MIT Press, 1969.

[8] Lo, Y.T., and S.W. Lee, (eds.), *Antenna Handbook: Theory, Applications, and Design,* New York: Van Nostrand Reinold, 1988, Chapter 7.

[9] King, R.W.P., "The Loop Antenna for Transmission and Reception." In *Antenna Theory (Part 1),* R.E. Collin and F.J. Zucker, (eds.), New York: McGraw-Hill, 1969.

[10] Kouyoumjian, R.G., "The Back-Scattering from a Circular Loop," *Appl. Sci. Res., Section B,* Vol. 6, 1956, pp. 165-179.

[11] Serdyukov, A.N., et al., *Electromagnetics of Bi-Anisotropic Materials: Theory and Applications,* Amsterdam: Gordon and Breach Science Publishers, 2001.

[12] Simovski, C.R., et al., "Antenna Model for Conductive Omega Particles," *J. Electromagnetic Waves and Applications,* Vol. 11, No. 11, 1997, pp. 1509-1530.

[13] Tretyakov, S.A., C.R. Simovski, and A.A. Sochava, "The Relation between Co-and Cross-Polarizabilities of Small Conductive Bi-Anisotropic Particles," *Advances in Complex Electromagnetic Materials,* Priou, A., et al., (eds.), NATO ASI Series High Technology, Vol. 28, pp. 271-280, Dordrecht, the Netherlands: Kluwer Academic Publishers, 1997.

[14] Landau, L.D., and E.M. Lifshits, *Electrodynamics of Continuous Media,* 2nd ed., Oxford, England: Pergamon Press, 1984.

[15] Sihvola, A., *Electromagnetic Mixing Formulas and Applications,* London: IEE Publishing, 1999.

[16] Vinogradov, A.P., *Electrodynamics of Composite Materials,* Moscow: URSS, 2001.

[17] Vinogradov, A.P., "On the Clausius-Mossotti-Lorenz-Lorentz Formula," *Physica A,* Vol. 241, 1997, pp. 216-222.

[18] Sihvola, A., "The Two Main Avenues Leading to the Maxwell Garnett Mixing Rule," *J. Electromagnetic Waves and Applications,* Vol. 15, No. 6, 2001, pp. 715-725.

[19] Sipe, J.E., and J. Van Kranendonk, "Macroscopic Electromagnetic Theory of Resonant Dielectrics," *Physical Review A,* Vol. 9, No. 5, 1974, pp. 1806-1822.

[20] Landauer, R., "Electrical Conductivity in Inhomogeneous Media," *Proc. of AIP Conference,* Vol. 40, New York: American Institute of Physics, 1978, pp. 2-45.

[21] Vinogradov, A.P., and A.M. Merzlikin, "On Electrodynamics of One-Dimensional Heterogeneous System Beyond Homogenization Approximation," *Advances in Electromagnetics of Complex Media and Metamaterials,* Zouhdi, S., A. Sihvola, and M. Arsalane, (eds.), Dordrecht, the Netherlands: Kluwer Academic Publishers, 2003, pp. 341-362.

[22] Vinogradov, A.P., and A.V. Aivaziyan, "Scaling Theory for Homogenization of the Maxwell Equations," *Physical Review E,* Vol. 60, 1999, pp. 987-993.

[23] Schwinger, J., et al., *Classical Electrodynamics,* Reading, MA: Perseus Books, 1998.

[24] Simovski, C.R., et al., "On the Surface Effect in Thin Molecular or Composite Layers," *European Physical J.,* Vol. AP 9, 2000, pp. 195-204.

[25] Mandelshtam, L.M., "Über optisch homogene und tübe Medien," *Annalen der Physik,* Bd. 23, 1907, S. 626-642 (Russian translation in *Complete Collection of Works,* Vol. 1, Moscow: USSR Academy of Science, 1957, pp. 109-170).

[26] Mandelshtam, L., "Zur Theorie der Dispersion," *Physikalische Zeitschrift,* Bd. 8, 1907, S. 608-611; Bd. 9, 1908, S. 308-311; S. 641-642 (Russian translation in *Complete Collection of Works,* Vol. 1, Moscow: USSR Academy of Science, 1957, pp. 162-172).

[27] Sobel'man, I.I., "On the Theory of Light Scattering in Gases," *Physics-Uspekhi*, Vol. 45, No. 1, 2002, pp. 75-80.

[28] Vinogradov, A.P., "On the Clausius-Mossotti-Lorentz-Lorenz Formula," *J. Communications Technology and Electronics*, Vol. 45, No. 8, 2000, pp. 811-819.

[29] Sivukhin, D.V., "Molecular Theory of Reflection and Refraction of Light," *Soviet Physics JETP*, Vol. 18, No. 11, 1948, pp. 976-994 (in Russian).

[30] Mahan, G.D., and G. Obermair, "Polaritons at Surfaces," *Physical Review*, Vol. 183, 1961, pp. 834-841.

[31] Collin, R.E., *Field Theory of Guided Waves*, 2nd ed., Piscataway, NJ: IEEE Press, and Oxford, England: Oxford University Press, 1991.

[32] Kontorovich, M.I., et al., *Electrodynamics of Grid Structures*, Moscow: Radio i Swiaz, 1987 (in Russian).

[33] Tretyakov, S.A., and A.H. Sihvola, "On the Homogenization of Thin Isotropic Layers," *IEEE Trans. Antennas and Propagation*, Vol. 48, No. 12, 2000, pp. 1858-1861.

[34] Collin, R.E., *Foundations for Microwave Engineering*, 2nd ed., Piscataway, NJ: IEEE Press, 2001.

[35] Ginzburg, V., *The Propagation of Electromagnetic Waves in Plasmas*, Oxford, England: Pergamon Press, 1964.

[36] Oksanen, M.I., J. Hänninen, and S.A. Tretyakov, "Vector Circuit Method for Calculating Reflection and Transmission of Electromagnetic Waves in Multilayer Chiral Structures," *IEE Proceedings-H*, Vol. 138, No. 6, 1991, pp. 513-520.

[37] Tretyakov, S.A., and M.I. Oksanen, "Electromagnetic Waves in Layered General Biisotropic Structures," *J. Electromagnetic Waves and Applications*, Vol. 6, No. 10, 1992, pp. 1393-1411.

[38] Belov, P.A., S.A. Tretyakov, and A.J. Viitanen, "Dispersion and Reflection Properties of Artificial Media Formed by Regular Lattices of Ideally Conducting Wires," *J. Electromagnetic Waves and Applications*, Vol. 16, No. 8, 2002, pp. 1153-1170.

[39] Brown, J., "Artificial Dielectrics," *Progress in Dielectrics*, Vol. 2, 1960, pp. 195-225.

[40] Rotman, W., "Plasma Simulation by Artificial Dielectrics and Parallel-Plate Media," *IRE Trans. Antennas and Propagation*, January 1962, pp. 82-95.

[41] Nicorovichi, N.A., R.C. McPhedran, and L.C. Botten, "Photonic Band Gaps for Arrays of Perfectly Conducting Cylinders," *Physical Review E,* Vol. 52, No. 1, 1995, pp. 1135-1145.

[42] Chin, S.K., N.A. Nicorovici, and R.C. McPhedran, "Green's Function and Lattice Sums for Electromagnetic Scattering by a Square Array of Cylinders," *Physical Review E,* Vol. 49, No. 5, 1994, pp. 4590-4602.

[43] Kuzmiak, V., A.A. Maradudin, and F. Pincemin, "Photonic Band Structures of Two-Dimensional Systems Containing Metallic Components," *Physical Review B,* Vol. 50, No. 23, 1994, pp. 16835-16844.

[44] Kuzmiak, V., A.A. Maradudin, and A.R. McGurn, "Photonic Band Structures of Two-Dimentional Systems Fabricated from Rods of a Cubic Polar Crystal," *Physical Review B,* Vol. 55, No. 7, 1997, pp. 4298-4311.

[45] Sakoda, K., et al., "Photonic Bands of Metallic Systems. I. Principle of Calculation and Accuracy," *Physical Review B,* Vol. 64, 2001, p. 045116.

[46] Ito, T., and K. Sakoda, "Photonic Bands of Metallic Systems. II. Features of Surface Plasmon Polaritons," *Physical Review B,* Vol. 64, 2001, p. 045117.

[47] Sigalas, M.M., et al., "Metallic Photonic Band-Gap Materials," *Physical Review B,* Vol. 52, No. 16, 1995, pp. 11744-11751.

[48] Sigalas, M., et al., "Photonic Band Gaps and Defects in Two Dimensions: Studies of the Transmission Coefficient," *Physical Review B,* Vol. 48, No. 19, 1993, pp. 14121-14126.

[49] Pendry, J.B., and A. MacKinnon, "Calculation of Photon Dispersion Relations," *Physical Review Lett.,* Vol. 69, No. 19, 1992, pp. 2772-2775.

[50] Guida, G., et al., "Mean-Field Theory of Two-Dimensional Metallic Photonic Crystals," *J. Optical Society of America B,* Vol. 15, No. 8, 1998, pp. 2308-2315.

[51] Simovski, C.R., M. Qiu, and S. He, "Averaged Field Approach for Obtaining the Band Structure of a Photonic Crystal with Conducting Inclusions," *J. Electromagnetic Waves and Applications,* Vol. 14, 2000, pp. 449-468.

[52] Felsen, L.B., and N. Marcuvitz, *Radiation and Scattering of Waves,* Piscataway, NJ: IEEE Press, 1991.

[53] Wait, J.R., "Reflection from a Wire Grid Parallel to a Conducting Plane," *Canadian J. Physics,* Vol. 32, 1954, pp. 571-579.

[54] Belov, P.A., et al., "Strong Spatial Dispersion in Wire Media at Extremely Low Frequencies," *Physical Review B,* Vol. 67, 2003, p. 113103.

[55] Shelby, R.A., D.R. Smith, and S. Schultz, "Experimental Verification of a Negative Index of Refraction," *Science,* Vol. 292, 2001, pp. 77-79.

[56] Maslovski, S.I., S.A. Tretyakov, and P.A. Belov, "Wire Media with Negative Effective Permittivity: A Quasi-Static Model," *Microwave and Optical Technology Lett.,* Vol. 35, No. 1, 2002, pp. 47-51.

[57] Contopanagos, H.F., C.A. Kyriazidou, and W.M. Merrill, "Effective Response Functions for Photonic Bandgap Materials," *J. Optical Society of America A,* Vol. 16, 1999, pp. 1682-1699.

[58] Moses, C.A., and N. Engheta, "An Idea for Electromagnetic 'Feedforward-Feedbackward' Media," *IEEE Trans. Antennas and Propagation,* Vol. 47, 1999, pp. 918-928.

[59] Tretyakov, S.A., and A.J. Viitanen, "Plane Waves in Regular Arrays of Dipole Scatterers and Effective-Medium Modeling," *J. Optical Society of America A,* Vol. 17, No. 10, 2000, pp. 1791-1797.

[60] Peltoniemi, J.I., "Variational Volume Integral Equation Method for Electromagnetic Scattering by Irregular Grains," *J. Quantum Spectroscopy and Radiation Transfer,* Vol. 55, 1996, pp. 637-647.

[61] Sihvola, A., and R. Sharma, "Scattering Corrections for the Maxwell Garnett Mixing Rule," *Microwave and Optical Technology Lett.,* Vol. 22, No. 4, 1999, pp. 229-231.

Chapter 6

Applications in Metamaterials and Artificial Impedance Surfaces

Definitions of the term *metamaterial* vary a lot (see a discussion in [1]), but its meaning is in fact rather close to that of *composite material*. *Meta* denotes position behind, after, or beyond, but also something of a higher or second-order kind [2]. This is the meaning we find in terms like *metalanguage* (a language used to describe another language). We will use the term *metamaterial* when speaking about media whose effective properties cannot be determined by only material parameters, shape, and concentration of the constituent inclusions. Usually this means that the inclusions are especially designed to provide the desired properties of the composite material. For example, we can fabricate inclusions of metal wires loaded by some electronic circuits that may include semiconductor devices such as transistors. Of course, these inclusions are composed of various usual materials (metals, dielectrics, semiconductors), but the resulting metamaterial properties clearly cannot be reduced to a kind of average of the constituent materials' properties. There is no well-defined boundary between composite materials and metamaterials in this loose understanding, especially when speaking about spatially dispersive media whose properties very much depend on the inclusion *shape* and on the electrical distance between inclusions, in addition to the size, concentration, and the inclusion material.

In this chapter we start by establishing the requirements for individual inclusions in a composite that should be satisfied to realize desired effective properties. We will consider mixtures of small inclusions, arrays of long conducting wires, and thin sheets of loaded conducting wires or patches.

6.1   METAMATERIALS WITH ENGINEERED AND ACTIVE INCLUSIONS

Radio engineers know and use artificial dielectrics, where molecules of usual electrically polarizable materials are replaced by electrically small but macroscopic scatterers, usually small metal spheres. This is simply a scaled model of normal dielectrics, whose main idea is that at low frequencies larger "molecules" are needed to get a strong enough electromagnetic response. Can the properties of these artificial molecules be adjusted for a specific application or made electrically tunable? In Chapter 5 we studied polarizabilities of short linear wire antennas and small loop antennas loaded by arbitrary bulk circuits. We saw that the polarizabilities can be tuned to the required values if the loads are appropriately chosen. This suggests that artificial materials with engineered properties can be designed from such artificial molecules if many loaded antennas are combined in a regular or random fashion.

6.1.1   Arrays of Small Loaded Wire Particles

Suppose that a material with a certain effective permittivity is required for a specific application. Let us try to design a metamaterial with this property using an array of small loaded dipole antennas. Each small wire dipole will imitate a molecule of a natural dielectric, and choosing the loads we try to synthesize the desired frequency dependence of the effective permittivity. In general, three arrays of antennas are needed, which should be arranged in space so that the required response is provided for electric fields along all directions. For the design of magnetic metamaterials, we also need loop antennas with appropriate loads (Figure 6.1).

Perhaps for the first time, similar complicated artificial media attracted interest with respect to the studies of chiral media for microwave applications.[1] Here, reciprocal magnetoelectric coupling can be realized in artificial composites with conductive particles or high-contrast dielectric inclusions of the helical shape. The canonical helix is a combination of a short dipole antenna and a small loop antenna connected so that one of the antennas is the load for the other. Analytical models for the polarizabilities of these loaded antennas were developed in [5]. These and similar models were used to study possible realizations of more complex bianisotropic [6,7], nonlinear [8], and other complicated metamaterials [9,10]. Here, we will first determine what polarizabilities of inclusions are needed if the desired permittivity and permeability are given.

---

[1] A review on analytical and numerical modeling of chiral and other bianisotropic particles can be found in [3,4].

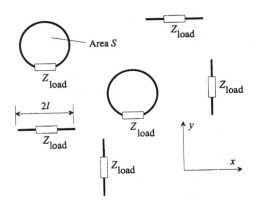

Figure 6.1 A spatial arrangement of small loaded wire and loop antennas. Particles in the $x - y$ plane are shown. Electric dipoles along x and y provide the required electric properties for electric fields in this plane. Loops give the desired response for magnetic fields along z. There are also two arrays of loops in the other two planes and an array of electric dipoles along z that are not seen here.

Required Polarizabilities and Loads

We will use the Clausius-Mossotti approach to model metamaterials built of small loaded antennas, assuming that the particle size is small compared to the wavelength, so that the inclusions can be modeled by electric or magnetic dipoles. More about the validity region of this approximation can be found in Section 5.2. The Clausius-Mossotti formula (5.81) shows that to realize a medium with the effective permittivity ϵ_{eff}, the inclusion electric polarizability α_{ee} should obey

$$N\alpha_{ee} = (\epsilon_{\text{eff}} - \epsilon_0)\frac{3\epsilon_0}{\epsilon_{\text{eff}} + 2\epsilon_0} \tag{6.1}$$

where N is the inclusion concentration and ϵ_0 is the matrix permittivity. From the duality principle it is obvious that the magnetic polarizability of inclusions α_{mm} for the design of a material with μ_{eff} is

$$N\alpha_{mm} = (\mu_{\text{eff}} - \mu_0)\frac{3\mu_0}{\mu_{\text{eff}} + 2\mu_0} \tag{6.2}$$

First we consider electric dipoles. Recall that the polarizabilities of short wire antennas with the input impedance Z_{inp} loaded by impedances Z_{load} read (5.17)

$$\alpha_{ee} = \left(\frac{4Z_{\text{inp}} + Z_{\text{load}}}{Z_{\text{inp}} + Z_{\text{load}}}\right)\frac{l^2}{3j\omega Z_{\text{inp}}} \tag{6.3}$$

Substituting into (6.1) and solving for the load impedance, we find what loads are required to realize effective permittivity ϵ_{eff}:

$$Z_{\text{load}} = \frac{(\epsilon_{\text{eff}} - \epsilon_0)\frac{3\epsilon_0}{\epsilon_{\text{eff}}+2\epsilon_0}Z_{\text{inp}} - \frac{4Nl^2}{3j\omega}}{\frac{Nl^2}{3j\omega Z_{\text{inp}}} - (\epsilon_{\text{eff}} - \epsilon_0)\frac{3\epsilon_0}{\epsilon_{\text{eff}}+2\epsilon_0}} \tag{6.4}$$

Let us compare the values of the two terms in the denominator of (6.4). The second term, $(\epsilon_{\text{eff}} - \epsilon_0)\frac{3\epsilon_0}{\epsilon_{\text{eff}}+2\epsilon_0}$, equals $N\alpha_{ee}$, see (5.81). The first term, $\frac{Nl^2}{3j\omega Z_{\text{inp}}}$, also contains most of the terms in the expression for $N\alpha_{ee}$ (5.17); only the resonant term $\frac{4Z_{\text{inp}}+Z_{\text{load}}}{Z_{\text{inp}}+Z_{\text{load}}}$ is missing. This means that in the resonant regime when this term is large in the absolute value, the second term in the denominator of (6.4) dominates over the first one, and the formula can be simplified as

$$Z_{\text{load}} = -Z_{\text{inp}} + \frac{4Nl^2(\epsilon_{\text{eff}} + 2\epsilon_0)}{9j\omega\epsilon_0(\epsilon_{\text{eff}} - \epsilon_0)} \tag{6.5}$$

We can view this impedance as that of a series connection of two loads. The impedance of the first one is the negative of the input impedance of the antenna Z_{inp}. The second one depends on the required effective permittivity. So, we can say that the frequency dispersion due to the shape and size of the metal wire is first compensated, and then the desired behavior is created by an additional load impedance.

The input impedance of a short wire dipole is capacitive, so we can substitute $Z_{\text{inp}} = 1/(j\omega C_{\text{wire}})$ in (6.4) and write[2]

$$Z_{\text{load}} = \frac{1}{j\omega}\frac{(\epsilon_{\text{eff}} - \epsilon_0)\frac{3\epsilon_0}{(\epsilon_{\text{eff}}+2\epsilon_0)C_{\text{wire}}} - \frac{4}{3}Nl^2}{\frac{Nl^2 C_{\text{wire}}}{3} - (\epsilon_{\text{eff}} - \epsilon_0)\frac{3\epsilon_0}{\epsilon_{\text{eff}}+2\epsilon_0}} \tag{6.6}$$

At this point we observe that for the design of "dispersion-free" lossless meta-materials (with a frequency independent and real effective permittivity), the load circuit should be a *capacitor*. The value of this required capacitance is obviously

$$C_{\text{load}} = \frac{\frac{Nl^2 C_{\text{wire}}}{3} - (\epsilon_{\text{eff}} - \epsilon_0)\frac{3\epsilon_0}{\epsilon_{\text{eff}}+2\epsilon_0}}{(\epsilon_{\text{eff}} - \epsilon_0)\frac{3\epsilon_0}{(\epsilon_{\text{eff}}+2\epsilon_0)C_{\text{wire}}} - \frac{4}{3}Nl^2} \tag{6.7}$$

For resonant inclusions, we have from (6.5)

$$\frac{1}{C_{\text{load}}} = -\frac{1}{C_{\text{wire}}} + \frac{4}{9}\frac{\epsilon_{\text{eff}} + 2\epsilon_0}{\epsilon_0(\epsilon_{\text{eff}} - \epsilon_0)}Nl^2 \tag{6.8}$$

Clearly, this value can be both positive and *negative*, depending on the required ϵ_{eff}. The last case can be realized only using active circuits.

---

[2]This result is applicable for regular lattices where the radiation resistance should be excluded from the input impedance; see Section 5.2.

To realize metamaterials with desired magnetic properties, we can use small loaded wire loops as artificial molecules. The magnetic polarizability of a small loaded loop is (5.46)

$$a_{mm} = -\mu_0^2 \frac{j\omega S^2}{Z_{\text{loop}} + Z_{\text{load}}} \qquad (6.9)$$

where S is the loop area. Substituting this in (6.2) and solving for Z_{load}, we find that to have the effective permeability of the metamaterial equal to μ_{eff}, the loops should be loaded by impedances

$$Z_{\text{load}} = -Z_{\text{loop}} - j\omega\mu_0 \frac{\mu_{\text{eff}} + 2\mu_0}{3(\mu_{\text{eff}} - \mu_0)} NS^2 \qquad (6.10)$$

Similar to the effective permittivity case, we see that the load impedance contains a term that compensates the input impedance of the loop antenna Z_{loop}. The input impedance of a small loop antenna is an inductance[3] $Z_{\text{loop}} = j\omega L_{\text{loop}}$, and if the engineered composite should be lossless (real μ_{eff}), the load impedance must be also an inductance:

$$L_{\text{load}} = -L_{\text{loop}} - \mu_0 \frac{\mu_{\text{eff}} + 2\mu_0}{3(\mu_{\text{eff}} - \mu_0)} NS^2 \qquad (6.11)$$

This "inductance" can take negative values: obviously, it must be negative to realize any frequency-independent real values of μ_{eff} that are larger than μ_0. Active circuits are necessary to design such loads (e.g., [11]).

6.1.2   Example: The Perfectly Matched Layer

The perfectly matched layer (PML) is a very effective method for the grid termination in finite difference techniques (see e.g., [12]). Because the PML has a very unique feature of zero reflection of arbitrarily polarized plane waves for arbitrary incidence angles, it appears to be of theoretical and practical interest to consider possibilities for a physical realization of this concept. The original formulation by Berenger [13] is based on introducing new field variables (splitting the physical fields into nonphysical components) that do not satisfy the usual Maxwell equations. This is not a problem for numerical implementations, but, obviously, no physical realization is possible. However, there exists a different formulation of the PML that can be expressed in terms of uniaxial material relations for the usual physical fields in an absorbing slab [14]. Consider a uniaxial magnetodielectric with the constitutive relations

$$\mathbf{D} = \epsilon_0(\epsilon_t \overline{\overline{I}}_t + \epsilon_n \mathbf{n}_0 \mathbf{n}_0) \cdot \mathbf{E}, \qquad \mathbf{B} = \mu_0(\mu_t \overline{\overline{I}}_t + \mu_n \mathbf{n}_0 \mathbf{n}_0) \cdot \mathbf{H} \qquad (6.12)$$

---

[3] As above for media with electric response, we again assume that the array of loaded loops is regular, so that the radiation resistance should not be included.

Here the unit vector $\mathbf{n}_0$ points in the direction orthogonal to the planar interface between free space and the medium (it also defines the only preferred direction in the structure). $\overline{\overline{I}}_t = \overline{\overline{I}} - \mathbf{n}_0\mathbf{n}_0$ is the transverse unit dyadic. The perfectly matched layer requirements for the relative material parameters read [14]:

$$\epsilon_t = \mu_t, \quad \epsilon_n = \mu_n, \quad \epsilon_n = \frac{1}{\epsilon_t}, \quad \mu_n = \frac{1}{\mu_t} \tag{6.13}$$

Indeed, inspecting (2.21) and (2.22) for the wave impedances of TE and TM polarized plane waves in uniaxial media, we see that if the relative material parameters satisfy (6.13), the wave impedances are the same as in free space, identically for all propagation directions (in other words, for all transverse wave vectors $\mathbf{k}_t$). Thus, an interface between free space and this medium is ideally matched for both polarizations and all incidence angles. In order to provide absorption, the transverse parameters ϵ_t, μ_t must be complex numbers with negative imaginary parts [remember that we work in the frequency domain, and the time dependence is $\exp(j\omega t)$]. Hence, the longitudinal components ϵ_n and μ_n have positive imaginary parts. The power absorbed or generated at a point inside a medium is (e.g., [15, Section 3.6])

$$\operatorname{Re}\{\nabla \cdot \mathbf{S}\} = \operatorname{Re}\left\{\frac{j\omega}{2}(\mathbf{E} \cdot \mathbf{D}^* - \mathbf{H}^* \cdot \mathbf{B})\right\} \tag{6.14}$$

In passive and absorbing media, this value must be negative *for all fields*:

$$\operatorname{Re}\{\nabla \cdot \mathbf{S}\} \leq 0 \tag{6.15}$$

In the uniaxial PML material this condition is not satisfied. For example, if at a certain point the electric field has only one component $\mathbf{E} = E\mathbf{n}_0$ and the magnetic field is zero,[4] the value of (6.14) is positive. This means that the medium is not passive: If at a certain point the transverse fields are small but the normal field components are large, energy must be pumped into the system at this point. Since in this medium *any individual* plane wave is absorbed, the conclusion that the medium is not passive can be confusing (see [16]). Actually, there is no contradiction: The power dissipated by a sum of two or more plane waves is not equal to the sum of powers dissipated by individual plane waves, as is obvious from (6.14).

Apparently, materials with electromagnetic properties defined by (6.12) and (6.13) are not found in nature, and the question arises if a material with such properties can be realized as a synthetic material. Similar results can be achieved with the use of higher-order spatial dispersion effects [17]. However, it is very difficult, if at all possible, to practically realize such materials (especially for operations in a wide frequency band). Indeed, the

---

[4]For example, in a standing wave formed by two plane waves traveling along the interface plane. In general, any field distribution can be realized by choosing the field sources.

material should very strongly interact with electromagnetic fields. That essentially means that the composite inclusions must be resonant particles, which means a narrow band of operational frequencies. This suggests the use of synthetic particles as described above, which can in principle provide the desired wideband response [9].

The results of Section 6.1.1 suggest that a possible composite material to realize the PML should contain six sets of different inclusions, in order to realize the four scalar material parameters in the constitutive relations. Electrically polarizable inclusions must be of two different types because the transverse and longitudinal properties are different, and we also need two types of magnetically polarizable particles. From the requirements on the PML material parameters we can find the requirements on the input impedances of the loads in order to provide the desired behavior of the composite. The solution for the load impedances is not unique, and we will consider only the simplest possible solutions. In all situations the required load impedance circuits are seen to be realizable as active electronic circuits, such as operational amplifiers with appropriate feedback circuits. The following is only an example of potential designs of composites with given properties. Many practical issues regarding the frequency band of possible operation and stability of the circuits need to be addressed for any practical realization.

Let us denote the loads of the four different sets of inclusions in the following way. Z_{wt} and Z_{wn} will stand for the loads of wire electric dipole antennas in the transverse plane (t) and directed along the normal to the surface (n), respectively. Z_{lt} and Z_{ln} are the loads for the loop particles. To create a material with the desired value of the transverse relative permittivity ϵ_t, the loads should be (6.5)

$$Z_{wt} = -Z_{\text{inp}} + \frac{4}{9j\omega\epsilon_0} \frac{\epsilon_t + 2}{(\epsilon_t - 1)} Nl^2 \tag{6.16}$$

The effective permittivity in the normal direction should equal $\epsilon_n = 1/\epsilon_t$; thus,

$$Z_{wn} = -Z_{\text{inp}} + \frac{4}{9j\omega\epsilon_0} \frac{1/\epsilon_t + 2}{(1/\epsilon_t - 1)} Nl^2 = -Z_{\text{inp}} - \frac{4}{9j\omega\epsilon_0} \frac{2\epsilon_t + 1}{(\epsilon_t - 1)} Nl^2 \tag{6.17}$$

The loads for the loop antennas are given by (6.10) after substitution of $\mu_t = \epsilon_t$ and $\mu_n = 1/\mu_t$:

$$Z_{lt} = -Z_{\text{loop}} - j\omega\mu_0 \frac{\epsilon_t + 2}{3(\epsilon_t - 1)} NS^2 \tag{6.18}$$

$$Z_{ln} = -Z_{\text{loop}} + j\omega\mu_0 \frac{2\epsilon_t + 1}{3(\epsilon_t - 1)} NS^2 \tag{6.19}$$

For the transverse fields the medium should be very lossy. Let us assume that $\epsilon_t = 1 - j\epsilon''$, $\epsilon'' > 0$ (the value of the real part is not important for the

PML performance, and with $\text{Re}\{\epsilon_t\} = 1$, formulas greatly simplify). With this choice we get

$$Z_{wt} = -\frac{1}{j\omega}\left(\frac{1}{C_{\text{wire}}} - \frac{4Nl^2}{9\epsilon_0}\right) + \frac{4Nl^2}{3\omega\epsilon_0\epsilon''} \qquad (6.20)$$

$$Z_{wn} = -\frac{1}{j\omega}\left(\frac{1}{C_{\text{wire}}} + \frac{8Nl^2}{9\epsilon_0}\right) - \frac{4Nl^2}{3\omega\epsilon_0\epsilon''} \qquad (6.21)$$

and

$$Z_{lt} = -j\omega\left(L + \mu_0\frac{NS^2}{3}\right) + \omega\mu_0\frac{NS^2}{\epsilon''} \qquad (6.22)$$

$$Z_{ln} = -j\omega\left(L - \mu_0\frac{2NS^2}{3}\right) - \omega\mu_0\frac{NS^2}{\epsilon''} \qquad (6.23)$$

The wire dipoles should be loaded by (negative) capacitances connected with resistances. The resistance is positive for the dipoles in the transverse plane and negative for the dipoles along the normal direction. The resistance depends on the frequency as $1/\omega$. Similar conclusions can be formulated for the loop particles, where the loads are negative inductances connected with positive or negative resistances. In this case resistances are directly proportional to the frequency.

This synthesis problem has infinitely many solutions. For example, we can demand the effective medium for the tangential fields behave as a conductor with a frequency independent conductivity σ, as in [16]:

$$\epsilon_t = 1 - j\frac{\sigma}{\omega\epsilon_0} \qquad (6.24)$$

This leads to the load impedances in the form

$$Z_{wt} = -\frac{1}{j\omega}\left(\frac{1}{C_{\text{wire}}} - \frac{4Nl^2}{9\epsilon_0}\right) + \frac{4Nl^2}{3\sigma} \qquad (6.25)$$

$$Z_{wn} = -\frac{1}{j\omega}\left(\frac{1}{C_{\text{wire}}} + \frac{8Nl^2}{9\epsilon_0}\right) - \frac{4Nl^2}{3\sigma} \qquad (6.26)$$

$$Z_{lt} = -j\omega\left(L + \mu_0\frac{NS^2}{3}\right) + \omega^2\frac{\epsilon_0\mu_0}{\sigma}NS^2 \qquad (6.27)$$

$$Z_{ln} = -j\omega\left(L - \mu_0\frac{2NS^2}{3}\right) - \omega^2\frac{\epsilon_0\mu_0}{\sigma}NS^2 \qquad (6.28)$$

Here, the load resistances for the dipole antennas do not depend on the frequency, but the resistive parts of the loop loads are proportional to the second power or ω. In principle, such loads can be realized using electronic circuits, but the problem is in potential instability of the system. Any plane electromagnetic wave will be absorbed, but more complicated field configurations can develop power generation because of the presence of negative resistances.

For many practical tasks it is most critical to design coverings that absorb energy of normally incident plane waves while absorption of obliquely incident waves is not so critical (e.g., antireflective coverings of aircrafts). In this situation the matched absorber requirement is only restricting the tangential components of the permittivity and permeability, so there is no problem with negative resistances needed for active inclusions of the PML. Let us study what loads are needed for our basic inclusions in a metamaterial in order to realize $\epsilon_t = \mu_t$. From (6.16) and (6.18), we find the required connection between the loads of the dipole and loop particles:

$$Z_{lt} + Z_{\text{loop}} = \frac{3S^2}{4l^2}\omega^2\epsilon_0\mu_0(Z_{wt} + Z_{\text{inp}}) \tag{6.29}$$

Substituting the input impedances of small particles $Z_{\text{loop}} = j\omega L$ and $Z_{\text{inp}} = 1/(j\omega C_{\text{wire}})$, we get

$$Z_{lt} = -j\omega\left(L + \frac{3S^2}{4l^2 C_{\text{wire}}}\epsilon_0\mu_0\right) + \frac{3S^2}{4l^2}\omega^2\epsilon_0\mu_0 Z_{wt} \tag{6.30}$$

In the same way we can express the required load of the electric dipole particles in terms of the load of the loops.

Since this is the only condition[5] imposed on two load impedances Z_{lt} and Z_{wt}, we can try, for instance, to choose such a load of the wire dipoles that would simplify the required load for the loops. Unfortunately, condition (6.30) cannot be satisfied with usual passive two-port circuits. This is mainly because the first term in this expression is a *negative* inductance. Indeed, replacing $j\omega$ by a complex variable p the matching condition (6.30) becomes

$$Z_{lt}(p) = -p\hat{L} - p^2 a Z_{wt}(p) \tag{6.31}$$

where $\hat{L} = L + \frac{3S^2}{4l^2 C_{\text{wire}}} > 0$ and $a = \frac{3S^2}{4l^2}\epsilon_0\mu_0 > 0$. If $Z_{wt}(p)$ is the input impedance of a passive circuit made up of bulk components, this is a ratio of two polynomials with real positive coefficients. As is obvious from (6.31), the coefficients of the polynomial on the left-hand side cannot be all positive. This means that only one of the two loads can be realized as a passive circuit of bulk components. In particular, we observe that if the wire antennas are loaded by a capacitance connected with a resistor, the loop inclusions should be loaded by a negative inductance and a resistor with a frequency depending resistance. Inversely, if the loops are loaded by a usual positive inductance and a normal resistor, the electric dipole inclusions should be loaded by a negative capacitance connected with a frequency-dependent resistor.

---

[5]Of course, there must be nonzero real part(s) of the load impedances to make the medium absorptive, which can be considered as another condition.

This discouraging result should be expected, because there is a fundamental limitation on the performance of passive absorbing layers [18]. If condition (6.30) would be possible to satisfy with bulk passive loads, then by choosing very high losses in the loads we would obtain a very thin absorbing layer matched in a very wide frequency range. Clearly, the use of active circuits in principle allows us to overcome this fundamental limit.

6.2 THIN ARTIFICIAL LAYERS AND SHEETS

Not only bulk materials can be artificially created, but also very thin layers or sheets. For example, we can simulate an electromagnetic response of a thin layer of a high-contrast dielectric or magnetic, or a layer with new properties can be created. First, let us discuss how a very thin resonant coating of a metal body can be designed [19].

6.2.1 Thin Resonant Coatings

Let us consider a regular planar square array with the period a formed by passive lossless scatterers located near an ideally conducting plane and parallel to that plane (Figure 6.2). The distance between the array plane and the ground plane is $h \ll \lambda$. If the array is nonresonant, the reflection coefficient for a normally incident plane electromagnetic wave (with the electric field amplitude $\mathbf{E}_{inc}$, the wave vector $\mathbf{k}$, and the frequency ω) from this structure is very close to -1, because the electric field exciting the dipoles is small.

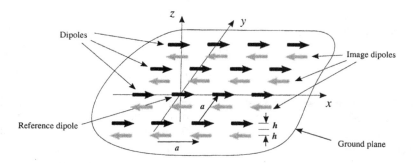

Figure 6.2 An array of small electric dipole particles located near a metal wall.

The interaction of the grid with the ideally conducting ground plane can be modeled using the image theory, replacing the ground plane by an array of image dipoles $\mathbf{p}' = -\mathbf{p}$. Let us choose the coordinate system with the zero-point in the center of one (reference) scatterer, with the z-axis normal to the ground plane and x- and y-axes parallel to the sides of the elementary cell as in Figure 6.2. In this coordinate system, the radius vectors of the centers of scatterers with numbers m, n are $\mathbf{R}_{m,n} = (m\mathbf{x} + n\mathbf{y})a$. The equation for the reference dipole moment can be written in the form:

$$\alpha^{-1}p = E_{\text{ext}} + E_{p'} + \sum_{(m,n)\neq(0,0)} [E_{p_{m,n}} + E_{p'_{m,n}}] \tag{6.32}$$

where p is the amplitude of the reference scatterer's dipole moment, E_{ext} is the external electric field, $E_{p'}$ is the field of the reference scatterer's image dipole, and $E_{p_{m,n}}$ and $E_{p'_{m,n}}$ are the fields of the other scatterers and their image dipoles, respectively. The external to the grid field is formed by the incident plane-wave field and the field of the incident wave reflected from the metal plane:

$$E_{\text{ext}} = 2j \sin{(kh)} E_{\text{inc}} \tag{6.33}$$

Expressing the electric field of the scatterers and the image dipoles via the dipole Green function $G(\omega, \mathbf{R})$

$$E_p(\mathbf{R}) = G(\omega, \mathbf{R})p \tag{6.34}$$

the equation for the unknown induced dipole moment (6.32) takes the form

$$\alpha^{-1}p = E_{\text{ext}} + \left\{ -G(2h\mathbf{z}_0) + \sum_{(m,n)\neq(0,0)} [G(\mathbf{R}_{m,n}) - G(\mathbf{R}_{m,n} + 2h\mathbf{z}_0)] \right\} p \tag{6.35}$$

where $\mathbf{z}_0$ is the unit vector along the z-axis. The coefficient in brackets is the interaction constant β:

$$\beta = -G(2h\mathbf{z}_0) + \sum_{(m,n)\neq(0,0)} [G(\mathbf{R}_{m,n}) - G(\mathbf{R}_{m,n} + 2h\mathbf{z}_0)] \tag{6.36}$$

Calculation of this function for planar dipole arrays has been explained in Sections 4.5 and 4.6. The induced dipole moment and the reflection coefficient can be now found in terms of the single particle polarizability α and the interaction constant β:

$$p = 2j \sin{(kh)} \left(\alpha^{-1} - \beta\right)^{-1} E_{\text{inc}} \tag{6.37}$$

$$R = -1 + \frac{k}{\epsilon_0 a^2} \frac{2j \sin^2{(kh)}}{\alpha^{-1} - \beta} = -1 + \frac{2}{1 - j\epsilon_0 a^2 \frac{\text{Re}\{\alpha^{-1} - \beta\}}{k \sin^2 kh}} \tag{6.38}$$

[compare with (4.136)]. Here, the first expression is general, and the second one is valid for lossless particles. Obviously, in the last case $|R| = 1$.

To realize the magnetic wall, the resonance condition should be satisfied: If $\text{Re}\{\alpha^{-1} - \beta\} \approx 0$, the reflection coefficient $R \approx +1$. The analysis shows [19], that usual particles like dielectric and metal spheres or small metal strips can never provide resonance response, basically because of small polarizability of these inclusions. Resonance can be achieved by loading particles, for example, short strip dipoles with capacitive input impedance, by bulk loads. If the load impedance is inductive, the particle polarizability near the particle resonance becomes large, which makes it possible to fulfill the resonance condition for a grid over the conducting plane.

Suppose that the array consists of short wire dipoles with total length $2l$ and radius r_0 loaded by bulk reactive impedances Z. For our purpose, the load impedance should be inductive, so we denote $Z = j\omega L$. Using the antenna model (Section 5.1) for determination of the polarizability of loaded wires, we have:

$$\alpha = \left(\frac{C_{\text{wire}}^{-1} - \omega^2 L}{l^2} + j\frac{k^3}{6\pi\epsilon_0} \right)^{-1} \qquad (6.39)$$

where $C_{\text{wire}} = \pi l \epsilon_0 / \log(2l/r_0)$ is the capacitance of the wire, L is the inductance of the load, l is the half-length of wire, and r_0 is the wire radius.

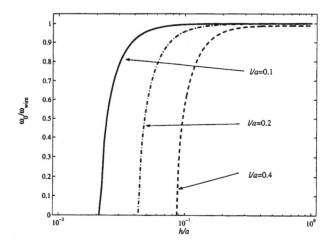

Figure 6.3 Dependence of the resonance frequency shift for reflection from an array of small loaded dipoles over a metal plane from the normalized thickness of the screen. In this example $l/r_0 = 10$.

We can see that indeed there is a resonance of the reflection coefficient with the central frequency

$$\omega_0 = \sqrt{\frac{C_{\text{wire}}^{-1} - \beta l^2}{L}} \qquad (6.40)$$

It is important to note that this value is smaller than the resonance frequency of individual loaded wires $\omega_{\text{wire}} = 1/\sqrt{LC_{\text{wire}}}$ due to the field interaction between inclusions. The dependence of the resonance frequency shift ω_0/ω_{wire} on the thickness of the screen [calculated using (6.40)] is plotted on Figure 6.3. When the height is large, the resonance is determined by the properties of the scatterer, with a small shift due to interactions within the array. At small heights, the field of the image dipole becomes large, and the resonance frequency of the reflection coefficient ω_0 becomes much smaller than the resonance frequency of a single array element ω_{wire}, and in theory the resonance frequency can approach zero. In practice this is restricted by the particle size.

6.2.2   Simulating Thin Material Layers

Thin layers of high-contrast dielectrics or magnetics can be modeled by sheet conditions (Chapter 2). The tangential electric and magnetic fields at the two sides of an equivalent infinitely thin sheet modeling a thin dielectric layer (relative permittivity ϵ_r, thickness d) are connected as (2.152)

$$\mathbf{E}_{t+} \approx \mathbf{E}_{t-}, \qquad \mathbf{n} \times \mathbf{H}_{t+} - \mathbf{n} \times \mathbf{H}_{t-} \approx \frac{j}{\eta_0}(\epsilon_r - 1)k_0 d\, \mathbf{E}_{t-} \qquad (6.41)$$

Because $\mathbf{E}_{t+} \approx \mathbf{E}_{t-}$, we can introduce the electric field in the sheet plane $\mathbf{E}_t = \mathbf{E}_{t+} = \mathbf{E}_{t-}$ and write simply

$$\mathbf{n} \times \mathbf{H}_{t+} - \mathbf{n} \times \mathbf{H}_{t-} = j\omega\epsilon_0 d(\epsilon_r - 1)\,\mathbf{E}_t \qquad (6.42)$$

For a layer of a magnetic material, we have (2.154)

$$\mathbf{H}_{t+} \approx \mathbf{H}_{t-} = \mathbf{H}_t, \qquad \mathbf{E}_{t+} - \mathbf{E}_{t-} = j\omega\mu_0 d(\mu_r - 1)\,\mathbf{n} \times \mathbf{H}_t \qquad (6.43)$$

In practice, it is often difficult if at all possible to find materials with the desired material parameters, which is especially true for magnetic materials at microwaves. Let us try to find a way to create layers with the same properties as material layers with given ϵ_r and μ_r with the use of metal inclusions and appropriate bulk loads. A natural design choice is the use of planar arrays of small metal wires or conducting patches that can be loaded by bulk impedances inserted into gaps between patches or as series loads of wires. This concept is illustrated by Figure 6.4: Here, an array of small conducting patches is loaded by lumped loads inserted between adjacent metal islands. Another possibility is to use a mesh of conducting wires or strips and load the wires by some bulk impedance circuits (Figure 6.5).

   To understand what kind of material layers can be simulated by these metalayers, we compare the boundary conditions for loaded arrays and thin material layers. Restricting the analysis to the normal incidence of plane waves, for an array of thin loaded strips[6] of width w, we have the boundary

---

[6]We consider planar strips in view of using duality with the patch arrays. For thin round wires the equations are very similar (see Chapter 4).

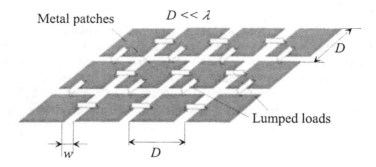

Figure 6.4 Small metal patches are connected by bulk inductances. Choosing the loads, it is possible to simulate various material layers. To provide isotropic response, a second set of loads is needed to connect the patch rows.

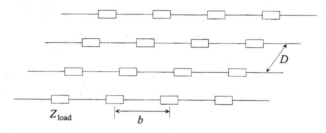

Figure 6.5 Wires forming a planar periodical array (period D) are periodically loaded by bulk loads (period $b \ll \lambda$) to change the grid properties.

condition (4.39)

$$\mathbf{E}_t = \left(ZD + j\frac{\eta}{2}\alpha\right)\mathbf{J} = \left(ZD + j\frac{\eta}{2}\alpha\right)(\mathbf{n}\times\mathbf{H}_{t+} - \mathbf{n}\times\mathbf{H}_{t-}) \qquad (6.44)$$

where the grid parameter is given by (4.56):

$$\alpha = \frac{kD}{\pi}\log\frac{1}{\sin\frac{\pi w}{2D}} \qquad (6.45)$$

and $Z = Z_{\text{load}}/b$. Equation (6.44) can be rewritten as

$$\mathbf{n}\times\mathbf{H}_{t+} - \mathbf{n}\times\mathbf{H}_{t-} = \frac{1}{ZD + j\omega\mu_0\frac{D}{2\pi}\log\frac{1}{\sin\frac{\pi w}{2D}}}\mathbf{E}_t \qquad (6.46)$$

Comparing this boundary condition for a loaded wire grid with the boundary condition for a dielectric layer (6.42), we can solve for the load impedance per unit length required to simulate a thin layer of dielectric with the relative permittivity ϵ_r:

$$Z = \frac{1}{jw\epsilon_0 dD(\epsilon_r - 1)} - jw\mu_0 \frac{1}{2\pi} \log \frac{1}{\sin \frac{\pi w}{2D}} \qquad (6.47)$$

In principle, we can substitute the desired frequency dispersive permittivity in (6.47) and find the required loads for the realization as a loaded wire mesh.

In (6.47), we again recognize a negative inductance term needed to compensate the surface inductance of the unloaded wire grid. This suggests that for simulating dielectric layers the use of capacitive patch arrays (Figure 6.4) is more appropriate, because its surface impedance is capacitive. For an unloaded array of small metal patches the boundary condition reads [see (4.59) and (4.60)]

$$\mathbf{n} \times \mathbf{H}_{t+} - \mathbf{n} \times \mathbf{H}_{t-} = jw\epsilon_0 \frac{2D}{\pi} \log \frac{1}{\sin \frac{\pi w}{2D}} \mathbf{E}_t = jwC_g \mathbf{E}_t \qquad (6.48)$$

where C_g is the grid capacitance per unit length. This is obviously of the same form as the boundary condition for a thin dielectric layer (6.42). If the array of patches is loaded by impedance per unit length Z, as shown in Figure 6.4, the total grid impedance is the parallel connection of $1/(jwC_g)$ and $Z_{\text{load}}D$. The load admittance needed to simulate a layer of a material with the relative permittivity ϵ_r is then

$$Y_{\text{load}} = jwD \left[\epsilon_0 d(\epsilon_r - 1) - C_g \right] \qquad (6.49)$$

As an example of potential applications, let us consider the use of such artificial layers as thin wideband absorbing coatings. Analyzing the formula for the reflection coefficient from a thin lossy dielectric layer on a metal surface at the normal incidence, we see (details in [4, Chapter 12]) that the condition on the dielectric permittivity needed to minimize reflection is

$$\epsilon_r \approx \frac{\pi^2}{4k_0^2 d^2} - j\frac{2}{k_0 d} \qquad (6.50)$$

where the product of the free-space wavenumber k_0 and the coating thickness d is supposed to be small: $k_0 d \ll 1$. Unfortunately, this dispersion rule is impossible in natural materials because of the Kramers-Kronig restriction [4]. From (6.49) we see that to simulate this behavior by an array of loaded patches we need the loads with the admittance

$$\frac{Y_{\text{load}}}{D} = j\frac{\pi^2}{4w\mu_0 d} - jwC_g + \frac{2}{\eta_0} \qquad (6.51)$$

This result has a clear physical meaning. The first term is a *negative* inductance needed to compensate the inductive part of the input impedance of a thin layer backed by a metal surface. The second term is a *negative* capacitance needed to compensate the surface capacitance of the patch array. Finally, the last term is a normal resistor needed to absorb the incident power. Obviously, impedance inverters are needed to realize such loads in practice.

Simulation of layers of magnetic materials requires arrays of particles possessing magnetic moments, such as spirals. This is obvious from the corresponding boundary condition (6.43): Because the tangential *electric* field is discontinuous, the surface must bear some *magnetic* current.

Finally, let us say that it is possible to solve for the required loads directly from requirements on the reflection and transmission coefficients, without considering any equivalent material layers. This can sometimes lead to simpler and more clear results.

6.3   LOADED WIRE MEDIA

Loads of individual particles like small electric dipoles and loops can be difficult to control, especially in three-dimensional lattices. Another possible technique to design engineered and controllable metamaterials can be based on the use of arrays of long conducting wires. The simplest "tuning" can be realized by periodically inserting semiconductor diodes in every wire, which can switch the current on or off [20]. The properties of these stopband materials (Chapter 5) can be tailored in more general fashion by periodically loading the wires by small impedance circuits. An application of arrays of capacitively loaded wires in the design of antenna reflectors was suggested in [21]. Load impedances can be electrically or optically controlled, thus allowing electrical control of the effective medium properties. A similar idea of loading wires by loops (inductive loads) was published in [22], where the goal was to reduce the effective plasma frequency of the medium. Periodical arrays of thin *loaded* wires can be analyzed by using the theory of arrays of thin ideally conducting cylinders (Section 5.5). The model presented here has been developed together with P.A. Belov and C.R. Simovski [23].

6.3.1   Dispersion Equation

Let us consider rectangular grids of infinite loaded wires as drawn in Figure 6.6. The elementary cell has dimensions $a \times b$. The radius of wires is $r_0 \ll a, b$, and they are periodically loaded by impedances $Z_{\text{load}}(\omega)$ [Ohm] with the period $c \ll a, b, \lambda$. In this situation, the loading can be interpreted as a uniformly distributed impedance $Z(\omega) = Z_{\text{load}}(\omega)/c$ [Ohm/m] per unit length of the wire. Here we will only study the mode with $E_z \neq 0$ that is affected by the wires.

Following the method in Section 5.5, we can calculate the local field acting at the surface of a reference wire and find the dispersion equation

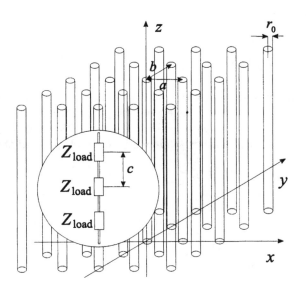

Figure 6.6 A periodical array of thin conducting periodically loaded cylinders. Loads Z_{load} can be arbitrary bulk two-port circuits.

for the eigenwaves in form (5.179). The calculation of the local field does not need any modifications, because the interaction constant is independent from the wire loads. The effective susceptibility of an ideally conducting continuous wire excited by a local electric field $\alpha_0 = I/E_z^{\text{loc}}$ is given by (5.167). If the wires are loaded by a uniformly distributed impedance per unit length $Z = Z_{\text{load}}/c$, the inverse wire susceptibility α becomes

$$\frac{1}{\alpha} = \frac{1}{\alpha_0} + \frac{k^2 - q_z^2}{k^2} Z \tag{6.52}$$

where α_0 is the susceptibility of an ideally conducting wire of the same radius. Here the coefficient $(k^2 - q_z^2)/k^2$ takes into account the influence of the local field phase shift along the wire. With this modification, the eigenvalue equation (5.179) becomes

$$\frac{1}{\pi} \log \frac{b}{2\pi r_0} + \frac{2}{j\eta k} Z + \frac{1}{bk_x^{(0)}} \frac{\sin k_x^{(0)} a}{\cos k_x^{(0)} a - \cos q_x a}$$

$$+ \sum_{n \neq 0} \left(\frac{1}{bk_x^{(n)}} \frac{\sin k_x^{(n)} a}{\cos k_x^{(n)} a - \cos q_x a} - \frac{1}{2\pi |n|} \right) = 0 \tag{6.53}$$

Here, as in Section 5.5, $k_x^{(n)}$ denotes the x-component of the wave vector of

nth Floquet mode:

$$k_x^{(n)} = -j\sqrt{\left(q_y + \frac{2\pi n}{b}\right)^2 + q_z^2 - k^2}, \qquad \text{Re}\{\sqrt{\cdot}\} > 0 \qquad (6.54)$$

For purely reactive loads Z, (6.53) is a real-valued dispersion equation whose solutions give dependencies of the eigenwave propagation constants q_x, q_y, q_z on the frequency ω. In the long-wavelength asymptotic solution (5.195), only the effective cutoff wavenumber k_p depends on the loads and the frequency:

$$q^2 = q_x^2 + q_y^2 + q_z^2 = k^2 - k_p^2 \qquad (6.55)$$

where

$$k_p^2 = \frac{2\pi/(ab)}{\log\dfrac{b}{2\pi r_0} + \dfrac{2\pi}{j\eta k}Z(k) + \displaystyle\sum_{n=1}^{+\infty}\left(\frac{\coth\frac{\pi n a}{b} - 1}{n}\right) + \dfrac{\pi a}{6b}} \qquad (6.56)$$

The assumptions used in the derivation of (6.55) are $q_x < \pi/a$, $q_y < \pi/b$, $q_z < \pi/c$. Note that it is not equivalent to the low frequency condition, because we can obtain rather high propagation constants in the regions of the impedance resonances.

6.3.2 Properties of Wire Media with Different Loads

Next we will give examples of effective properties that can be realized using wire arrays loaded by various simple reactive loads. We start from inductive loadings.

Inductively Loaded Wire Media

Considering the reflection coefficient from a half-space filled by an array of ideally conducting wires (Section 5.5, Figure 5.19), we can note that near the upper edge of the low-frequency stopband the interface between free space and the wire medium operates as a magnetic wall. This property can be very useful in antenna applications, because a wire antenna placed over a magnetic screen does not suffer destructive influence from the current of the screen, but instead experiences double amplification of the radiated field. The frequency of the upper edge gets smaller if the wire radius is reduced, but the manufacturing restrictions and the skin effect do not allow us to obtain this edge at relatively low frequencies (such as required for most antenna applications). As we see from (6.55) and (6.56), inductive loads $Z = j\omega L$ connected in series with the wire sections reduce the value of k_p, thus making the low frequency bandgap narrower [22] and shifting the magnetic-wall regime to considerably lower frequencies. This is illustrated

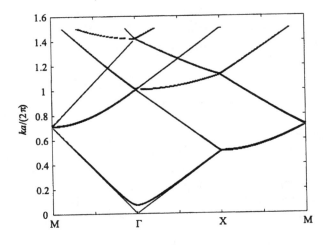

Figure 6.7 Dispersion plot for a square grid of cylinders with the filling ratio $f = 0.001$ loaded by inductive impedance per unit length $L = 5\mu_0$.

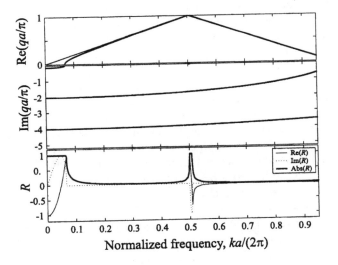

Figure 6.8 Dispersion curves and the reflection coefficient from a half-space for the same array as in Figure 6.7.

by Figures 6.7 and 6.8, which show the dispersion curves and the reflection coefficient for an inductively loaded wire array. Numerical estimations show that it is realistic to obtain an artificial magnetic wall at such low frequencies that the array periods a and b are only small fractions of the wavelength. In this example, the array period $a < 0.1\lambda$ in the magnetic-wall regime. Inductive loads can also be used to create filters with a controllable frequency band.

Capacitively Loaded Wire Media

Next, let us consider arrays of wires loaded by capacitances. This means that the wires are periodically cut (the distances between the cuts are much smaller than the wavelength) and a bulk capacitance inserted into every gap. Every wire can be seen as made up of series connections of these load capacitances and inductances formed by wire sections between the loads. The effective medium behavior dramatically depends on the resonant frequency of these sections. If the load capacitance tends to infinity or the frequency is very high (this corresponds to unloaded cylinders, because the impedance of the loads tends to zero) the system behaves like an array of thin ideally conducting cylinders (Section 5.5). In the case when the capacitance is infinitely small (which corresponds to interrupted wires), the medium behaves like a three-dimensional lattice of discrete scatterers (Section 5.6).

The dispersion plot, reflection coefficient from a half-space, and the corresponding propagation constants for a square grid of cylinders with the filling ratio $f = 0.001$ loaded by capacitive impedances with a certain capacitance value (capacitance per unit length in this example is $C = 2\epsilon_0 a^2$) are presented in Figures 6.9 and 6.10. The topology of the propagation constant plot looks similar to the unloaded case except for the appearance of a low-frequency passband and exponentially decaying modes with alternating current directions $\mathrm{Re}(q) = \pi/a$ existing at frequencies higher than the upper edge of that passband (Figure 6.10). The reflection coefficient from a half-space at frequencies within the first stopband is mainly determined by the evanescent modes with the smallest decay factors, and in this particular case we observe that inside the first stopband near the lower band edge, the mode with $\mathrm{Re}(q) = \pi/a$ has a smaller decay factor, but at higher frequencies up to the upper band edge, the mode with $\mathrm{Re}(q) = 0$ determines the reflection properties. Similar effects inside stopbands produced by resonances of inclusions were observed in three-dimensional lattices of resonant ferrite spheres [24].

If the self-resonance frequency of loaded wires is higher than the frequency of the first lattice spatial resonance (practically meaning that the load capacitance is small), the low-frequency bandgap completely disappears, and the first branch of the dispersion curves takes the same form as for a three-dimensional lattice of small scatterers.

Series LC-circuit loads operate in the same way as the capacitive loads in wire media with the wire radius effectively reduced by inductive loading; see the discussion on inductive loads above.

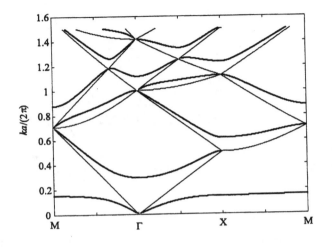

Figure 6.9 Dispersion plot for a square grid of cylinders with the filling ratio $f = 0.001$ loaded by capacitive impedances with $C = 2\epsilon_0 a^2$ per unit length.

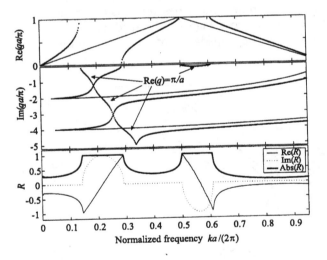

Figure 6.10 Reflection coefficient (at normal incidence and for the electric field polarized along the wires) from a half-space filled by the same grid as in Figure 6.9 and the corresponding propagation constants.

In the quasistatic regime for fields independent of z (i.e., $q_z = 0$), in the case of capacitive $Z = 1/(j\omega C)$ and series LC-circuit $Z = 1/(j\omega C) + j\omega L$

loads, we have a resonant effective permittivity in the form:

$$\epsilon(\omega) = \epsilon_0 \left[1 + \frac{C/(\epsilon_0 ab)}{1 - \omega^2/\omega_p^2} \right] \tag{6.57}$$

where

$$\omega_p^2 \doteq \frac{2\pi/(\mu_0 C)}{\log\frac{b}{2\pi r_0} + \frac{2\pi L}{\mu_0} + \sum\limits_{n=1}^{+\infty} \left(\frac{\coth\frac{\pi n a}{b} - 1}{n} \right) + \frac{\pi a}{6b}} \tag{6.58}$$

At frequencies lower than the circuit resonance the medium operates as an artificial dielectric, and at higher frequencies the medium becomes an artificial plasma.

Parallel Resonant LC-Circuit Loading

Combinations of inductive and capacitive loads connected in parallel give us an ability to control the medium dispersion in a more general fashion. If the lumped loads are parallel LC-circuits, the main resulting effect is seen in the appearance of a transparency band near the resonance frequency of the circuit and a stopband at higher frequencies (Figures 6.11 and 6.12). A passband is formed around the series resonance frequency, where the wires are weakly excited due to a high total impedance of wires per unit length. At high frequencies the impedance of the loads is very small, so they do

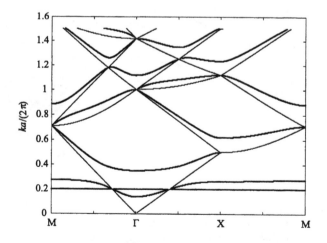

Figure 6.11 Dispersion plot for a square grid of cylinders with the filling ratio $f = 0.001$ loaded by a parallel resonant circuit with $L = 0.5\mu_0$ inductance tuned to the resonant frequency corresponding to the wavenumber $k_{\text{res}}a/(2\pi) = 0.2$ (shown by a horizontal line).

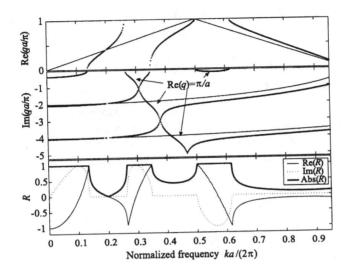

Figure 6.12 Reflection coefficient (at normal incidence and for the electric field polarized along the wires) from a half-space filled by the same medium as in Figure 6.11 and the corresponding propagation constants.

not influence the array properties. Inside the new passband of the medium, the reflection coefficient from a half-space changes from plus one (the lower end) to minus one (the upper end), passing through zero at the center of the passband, where the medium becomes transparent.

We can position the self-resonance of the load circuit at the required frequency and design electromagnetic crystals with the desired band structure. Furthermore, the bandgap structure in this case is modified only near the load resonance. Far from that frequency the waves are naturally not affected by the loads. Thus, for example, we can tune the structure so that there is a very narrow passband inside the wide low-frequency stopband. Also, it is possible to have a narrow stopband inside a wide passband of the structure, increasing the resonance frequency of the loads. In the case of parallel LC-circuit loads $Z(\omega) = j\omega L/(1 - \omega^2 LC)$, the effective permittivity (again, introduced for $q_z = 0$) tends to infinity and also passes through unity at nearly positioned frequencies:

$$\epsilon(\omega) = \epsilon_0 \left(1 - \frac{2\pi/(\epsilon_0 \mu_0 \omega^2 ab)}{\log \frac{b}{2\pi r_0} + \frac{2\pi L/\mu_0}{1 - \omega^2 LC} + \sum_{n=1}^{+\infty} \left(\frac{\coth \frac{\pi n a}{b} - 1}{n} \right) + \frac{\pi a}{6b}} \right) \quad (6.59)$$

At the circuit resonant frequency the value of the load becomes infinite, and the medium becomes transparent with $\epsilon = \epsilon_0$. At higher frequencies the load behaves like a capacitance, and there is also a resonance of the medium, where the effective permittivity in the lossless case becomes very large.

6.3.3   Quasistatic Modeling of Arrays of Lossy and Loaded Wires

The quasistatic model for this system was developed in Section 5.5.4, but only for ideally conducting wires. If the wires can be described by some effective surface impedance $Z = Z(\omega)$ in addition to the inductance, the following formula for the permittivity can be obtained in a similar way [25]:

$$\overline{\overline{\epsilon}} = \epsilon_0 \left[\overline{\overline{I}} - \frac{2\pi \mathbf{z}_0 \mathbf{z}_0}{(ka)^2 \log \frac{a^2}{4r_0(a-r_0)} - jka\frac{a}{r_0}\frac{Z}{\eta}} \right] \tag{6.60}$$

where $\eta = \sqrt{\mu_0/\epsilon_0}$ is the free-space impedance. For example, for lossy wires (skin depth much smaller than the wire diameter)

$$Z = \frac{(1+j)}{\sqrt{2}} \sqrt{\frac{\omega\mu_0}{\sigma}} \tag{6.61}$$

where σ is the metal conductivity, and we see that if the wires are lossy, the real part of the permittivity (which is negative) decreases in the absolute value and the permittivity naturally becomes complex. Note that the loss effect increases at low frequencies, because the main term in the denominator of (6.60) decreases as ω^2, but the loss term decreases as $\omega^{3/2}$. Losses can be neglected if

$$\frac{|Z|}{\eta} \ll kr_0 \log \frac{a^2}{4r_0(a - r_0)} \tag{6.62}$$

In a situation when the skin effect can be neglected (e.g., very low frequencies), term $Z/(2\pi r_0)$ in (6.60) should be replaced by the static resistance of wire per unit length $1/(\pi r_0^2 \sigma)$. The effective permittivity reads in this case

$$\overline{\overline{\epsilon}} = \epsilon_0 \left[\overline{\overline{I}} - \frac{2\pi \mathbf{z}_0 \mathbf{z}_0}{(ka)^2 \log \frac{a^2}{4r_0(a-r_0)} - 2jk\frac{a^2}{r_0^2}\frac{1}{\sigma\eta}} \right] \tag{6.63}$$

In this situation losses can be neglected if

$$\frac{1}{\pi r_0^2 \sigma \eta} \lambda \ll \log \frac{a^2}{4r_0(a - r_0)} \tag{6.64}$$

The expression on the left side is the ohmic resistance of a one-wavelength-long piece of wire normalized to the free-space impedance η.

With this model, we can also estimate the low-frequency behavior of more complicated and exotic metamaterials that can be realized by loading wires by reactive impedances. For example, if we periodically cut the wires (and possibly insert bulk capacitances into every cut), (5.203) becomes

$$E_z = I \left(j\omega L + \frac{1}{j\omega C} \right) \qquad (6.65)$$

from where we find the effective permittivity as

$$\overline{\overline{\epsilon}} = \epsilon_0 \left[\overline{\overline{I}} + \frac{\mathbf{z}_0 \mathbf{z}_0}{j\omega \epsilon_0 a^2 \left(j\omega L + \frac{1}{j\omega C} \right)} \right] = \epsilon_0 \left[\overline{\overline{I}} + \frac{\mathbf{z}_0 \mathbf{z}_0 \, C}{\epsilon_0 a^2 (1 - \omega^2 LC)} \right] \qquad (6.66)$$

In (6.65) the expression in brackets is the total wire impedance per unit length. Thus, capacitance C is measured in F·m.

At low frequencies, we have an effective medium with a positive permittivity defined by the capacitances of the gaps. Well above the resonance, the material has negative permittivity for electric fields directed along the wires. For high capacitive impedances, (5.203) becomes

$$E_z = I \left(j\omega L + \frac{1}{j\omega C} \right) \approx I \frac{1}{j\omega C} \qquad (6.67)$$

Substituting this in (5.205), we note that the frequency dependence cancels out, and we get

$$\mathbf{D} = \left(\epsilon_0 \overline{\overline{I}} + \mathbf{z}_0 \mathbf{z}_0 \frac{C}{a^2} \right) \cdot \mathbf{E} \qquad (6.68)$$

corresponding to a simple artificial dielectric.

6.4   ARTIFICIAL IMPEDANCE SURFACES

In some antenna and waveguide applications it is desirable to have surfaces characterized by high reactive impedances. For example, to design low-profile wire antennas, the radiating wire is positioned close to the supporting body (of an aircraft or a car, for example). This situation is illustrated in Figure 6.13. Obviously, if the ground plane is an ideal conductor, the radiation is very poor because of the far-zone cancellation of the wire field and the field created by its mirror image in the plane. Actually, this is not an antenna but a good transmission line. The situation reverses if the ground plane is a magnetic conductor, meaning that on its boundary the tangential component of *magnetic* field vanishes. The mirror image current changes its sign, and the field radiated by the wire in space is doubled with respect to the field of the same wire in free space. If the distance between the wire and the ground plane is small, and the ground plane is a magnetic conductor, then looking from a distance the antenna appears to radiate as a single line of current. Further in this chapter we will see that this can be a disadvantage of magnetic conductors as ground planes. The main reason is that in realistic applications the size of the ground plane is limited, and it is often desirable to have small fields and currents at the edges of the ground plane. In the usual case of an ideally conducting ground plane the radiation pattern has a null in the direction along the surface.

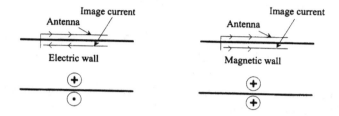

Figure 6.13 Wire antennas near ideal electric and magnetic surfaces. The mirror image of the antenna current in magnetic wall is in phase with the wire current.

There is of course no problem to realize electric walls. They are surfaces of good conductors, which allow electric charges to freely flow on the surface. Magnetic walls would require good conductors for magnetic charge, but unfortunately there are no magnetic charges in nature. So how can we realize magnetic conductors? Consider plane-wave reflection from a boundary characterized by a scalar surface impedance Z_s. For the normal incidence of plane waves, the reflection coefficient for the electric field is

$$R_e = \frac{Z_s - \eta}{Z_s + \eta} \tag{6.69}$$

(η is the wave impedance). As we all know, the electric wall corresponds to $Z_s = 0$, which gives $R_e = -1$, that is, zero tangential electric field on the interface. In the case of a magnetic wall we should have $R_e = +1$, which corresponds to zero tangential magnetic field. Clearly, this is realized when $|Z_s| \to \infty$. This brings us to the concept of *artificial high-impedance surfaces*: They are characterized by high absolute values of the surface impedances and can be used to realize magnetic walls. The next question is how to design structures with high surface impedances.

Let us simply cover a metal plane with a magnetodielectric layer of thickness d (Figure 6.14). The input impedance seen at the upper surface by normally incident plane waves is

$$Z_s = j\sqrt{\frac{\mu}{\epsilon}} \tan(\omega\sqrt{\epsilon\mu}d) = j\sqrt{\frac{\mu}{\epsilon}} \tan\left(\frac{2\pi}{\lambda}d\right) \tag{6.70}$$

At $d = \lambda/4$ this tends to $j\infty$, realizing a magnetic wall. In practice, this principle is realized as metal plates with periodical grooves (*corrugated* surfaces).

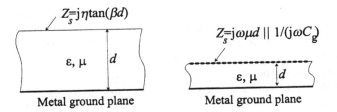

Figure 6.14 Realization of a magnetic wall as a dielectric cover on a metal wall and as an artificial high-impedance surface. In the second case the thickness can be greatly reduced, but the operational bandwidth decreases.

When the period of corrugations is small compared to the wavelength, the surface can be described by an equivalent surface impedance that connects the tangential components of the averaged electric and magnetic fields on the surface. If the depth of the grooves is at resonance with the wavelength of the exciting electromagnetic field, the impedance values can be rather high. The biggest problem is that for practical applications a quarter-wavelength cover is prohibitively thick.

To reduce the thickness, we can use the system shown on the right of Figure 6.14. Here, the thickness of the layer is much smaller than the wavelength, so the input impedance on the layer surface is inductive (and quite small in the absolute value): $Z_{\text{inp}} \approx j\omega\mu d$. The main idea is to position a *capacitive* grid (e.g., an array of small metal patches) on the surface of the layer. The capacitive grid can be modeled by a capacitive grid impedance $Z_g = 1/(j\omega C_g)$ (see Chapter 4). The total surface impedance is the parallel connection of the inductive input impedance of the thin layer and the capacitive grid impedance of the array:

$$Z_s = \frac{j\omega\mu d}{1 - \omega^2 C_g \mu d} \qquad (6.71)$$

At the resonance frequency of the system $\omega_0 = \sqrt{1/(C_g \mu d)}$, the imaginary part of the surface impedance tends to infinity and the system is a magnetic wall (again, for normally incident plane waves). Such composite layers as artificial impedance surfaces were suggested by D. Sivenpiper et al. [26, 27], and they are called *mushroom structures*. In that design, every patch of the capacitive array is connected to the ground by a thin vias (Figure 6.15). Vias connectors are important for obliquely incident waves, and we will discuss that later.

High impedance can be also realized using planar resonant grids. We saw such behavior in simple arrays of conducting strips with rectangular cells (Section 4.4.2, Figure 4.9). Here, the inductive impedance of the strips is connected in parallel with effective capacitances between the strips in the orthogonal array. At the resonant frequency the total grid impedance becomes very high. The performance can be improved by varying geometry of the cell (in particular, capacitance can be made larger). This solution was introduced by the research group lead by T. Itoh [28], who suggested naming such resonant grids *uniplanar high-impedance surfaces*. Note that there is a fundamental difference between high-impedance surfaces that use patch arrays over a metal plane and resonant grids. In the last case, it is the grid impedance that takes large imaginary values at the resonance, but not the surface impedance. This means that at the resonance the grid actually becomes transparent, in contrast to arrays over a conducting plane.

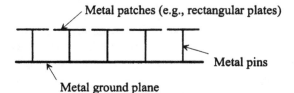

Figure 6.15 Patches can be grounded by vias wires to prevent waves with nonzero vertical electric field component to travel between the patch array and the ground.

The advantages of composite materials in the design of high-impedance surfaces can be derived from the fact that some quasi-bulk elements are used, and the field is concentrated in some small volumes in or near the layer. Sometimes these structures are called two-dimensional photonic bandgap structures because of their two-dimensional periodicity and the presence of a stopband for surface waves. Potential applications are in antennas [26, 29] and microwave filters [28].

In modeling both planar and three-dimensional structures, local quasistatic models were used in [26–28]. In the simple circuit model of mushroom layers introduced by Sievenpiper et al., the structure is essentially considered as an array of noninteracting cells, although they share the same magnetic flux. Each cell is characterized by its quasistatic parameters: capacitance and inductance. The equivalent capacitance is calculated as the capacitance of a single cell, that between two adjacent patches. Thus, no field interaction between cells through their electric fields is taken into account. This local and quasistatic approximation results in an equivalent circuit representation of a unit cell in the form of a parallel circuit. However, we know from Chapter 4 that patches in dense arrays strongly interact, and the parameters of arrays can be only roughly estimated from the parameters of individual inclusions. A more accurate model was used in [30] in studies of TM waves along the Sievenpiper mushroom surface. In that study, the space between the patches and the ground was replaced by a transmission line filled by a uniaxial dielectric, and the patch array was modeled by an equivalent capacitive reactance.

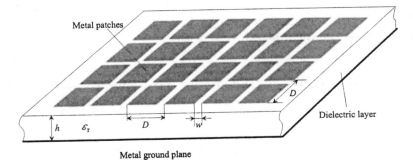

Figure 6.16 Geometry of an artificial impedance surface. An array of conducting patches is located at distance h from the ground plane. The space between the array and the ground is filled by a dielectric with the relative permittivity ϵ_r.

We will start from building a dynamic model of three-dimensional (mushroom) artificial surfaces, which takes into account electromagnetic interactions of all cells.[7] The geometry can be seen from Figure 6.16. The present model is valid for small square patches compared to the wavelength (array period $D \ll \lambda$) and narrow slots between patches (slot width $w \ll D$). The distance to the ground is h. This system can be analyzed numerically, using for example, the periodical method of moments, but our goal here is to develop an analytical full-wave model.

First, we consider the normal-incidence plane-wave excitation. In particular, this means that the vias conductors connecting the patches to the ground are not excited, and there is no need to take them into account. So the system we model consists of a periodical planar array of ideally conducting patches positioned parallel to an infinite conducting plane, at a small distance. A dielectric substrate supports the patch array.

6.4.1 Equivalent Surface Impedance of Mushroom Structures

Here, we introduce and calculate the equivalent surface impedance of reactive impedance surfaces. We will do it for both capacitive grids near conducting planes (arrays of patches as in Figure 6.16) and the complementary structure of inductive grids (arrays of thin conducting strips in place of the gaps between the patches).

---

[7]This model was developed together with C.R. Simovski [31].

Grid Impedance

Electromagnetic properties of planar grids of this kind can be described in terms of the grid impedance Z_g, which connects the averaged electric field in the grid plane and the averaged current density:

$$\widehat{E} = Z_g \widehat{J} \tag{6.72}$$

In the case of an inductive grid formed by thin parallel conducting strips of width w, the grid impedance in (6.72) is given by (4.39):

$$Z_g = Z_{\text{strips}} = j\frac{\eta}{2}\alpha \tag{6.73}$$

where

$$\alpha = \frac{kD}{\pi} \log\left(\frac{2D}{\pi w}\right) \tag{6.74}$$

is the grid parameter (4.57), $k = \omega\sqrt{\epsilon_0\mu_0}$, and the wave impedance $\eta = \sqrt{\mu_0/\epsilon_0}$. This relation is suitable for thin conducting strips as compared to the grid period ($w \ll D$). More general results are given in Chapter 4. The averaging and homogenization procedures leading to (6.72) take into account electromagnetic interactions in infinite grids. Simplifications are in the cell models: The cell size is assumed to be small compared to the wavelength, so that the local field distribution over a cell is assumed to be close to the quasistatic distribution.

For the complementary array of conducting patches in free space, the grid impedance can be found using the Babinet principle (e.g., [32]):

$$Z_g = Z_{\text{patches}} = \frac{\eta^2}{4Z_{\text{strips}}} \tag{6.75}$$

In terms of the grid parameter α, this gives

$$Z_{\text{patches}} = -j\frac{\eta}{2\alpha} \tag{6.76}$$

that is, simply replace α by $-1/\alpha$ in (6.73). If there is a dielectric material on one side of the grid (and there is free space on the other side), the symmetry that leads to the Babinet principle is missing, and (6.75) is not valid. However, there exists an approximate formula [33]

$$Z_{\text{grid}} = Z_{\text{patches}} = \frac{\eta^2}{4Z_{\text{strips}}}\frac{2}{\epsilon_r + 1} = -j\frac{\eta}{2}\frac{1}{\alpha}\frac{2}{\epsilon_r + 1} \tag{6.77}$$

where ϵ_r is the relative permittivity of the lower half-space. The last formula is very accurate for grids whose period is small compared to the wavelength, which is our case of interest.

This theory neglects electric polarization in thin conducting strips ($w \ll D$) when the incident electric field is orthogonal to the strip. An approximate formulation of the Babinet principle that takes into account that effect has been published in [34].

Equivalent Surface Impedance and Reflection Coefficient

Let us now consider a grid positioned parallel to a conducting plane, and assume first that the distance from the grid to the ground h is not smaller than the grid period D. In this situation we can neglect higher-order Floquet modes generated by the periodical mesh. Assuming only the fundamental-mode plane waves between the array and the ground, the equivalent surface impedance can be easily found as the impedance of a parallel connection of the grid impedance Z_g and the input impedance of a TEM line section of length h ($Z = j\eta_- \tan(k_- h)$, where $k_- = k\sqrt{\epsilon_r}$ and $\eta_- = \eta/\sqrt{\epsilon_r}$ are the parameters of the medium between the array and the ground plane). The input impedance is found from

$$\frac{1}{Z_s} = \frac{1}{j\eta_- \tan(k_- h)} + \frac{1}{Z_g} \tag{6.78}$$

that is,

$$Z_s = \frac{Z_g \tan(k_- h)}{\tan(k_- h) - j\frac{Z_g}{\eta_-}} \tag{6.79}$$

Interpretation of Z_s singularity at the frequency where the denominator of (6.79) equals zero as a parallel resonance of the array and the transmission line formed by spacing h between the patches and the ground plane is evident.

In terms of the grid parameter α, we have

$$Z_s = \frac{j\eta\frac{\alpha}{2} \tan(k_- h)}{\tan(k_- h) + \sqrt{\epsilon_r}\frac{\alpha}{2}} \tag{6.80}$$

for arrays of conducting strips and, using (6.77),

$$Z_s = \frac{j\frac{\eta}{\sqrt{\epsilon_r}} \tan(k_- h)}{1 - \frac{\epsilon_r+1}{\sqrt{\epsilon_r}}\alpha \tan(k_- h)} = \frac{j\frac{\eta}{\sqrt{\epsilon_r}} \tan(k_- h)}{1 - \frac{(\epsilon_r+1)kD}{\pi\sqrt{\epsilon_r}} \log\left(\frac{2D}{\pi w}\right) \tan(k_- h)} \tag{6.81}$$

for arrays of patches. The last simple formula can be used to calculate the equivalent surface impedance of Sievenpiper impedance surfaces for normally incident plane waves. As will be shown in the next section, the result can be extended to resonant grids and corrected to account for higher-order Floquet modes by appropriate modifications of the grid parameter α in (6.80).

Replacing $\tan(k_- h)$ by its argument for the case of small $|k_-|h$ yields Z_s in the form that coincides with the result from the parallel circuit model [26]:

$$Z_s = \frac{j\omega L}{1 - \omega^2 LC} \tag{6.82}$$

where

$$L = \mu_0 h, \qquad C = \frac{D\epsilon_0(\epsilon_r + 1)}{\pi} \log\left(\frac{2D}{\pi w}\right) \tag{6.83}$$

It is important that for realistic and practical sizes the resonant frequency is not very low due to rather small values of the effective capacitance and inductance, and the structure thickness is not negligible as compared to the resonant wavelength. Therefore, the use of the formula for the resonant frequency that follows from (6.82) (i.e., $\omega_0 = 1/\sqrt{LC}$), leads to considerable errors, and it is preferable to use (6.81).

The reflection coefficient from the surface is, obviously,

$$R = \frac{Z_s - \eta}{Z_s + \eta} \tag{6.84}$$

If there is no dielectric layer between the array and the ground plane ($\epsilon_r = 1$), the reflection coefficient can also be written as

$$R = -e^{-2jkh} - \frac{\left(1 - e^{-2jkh}\right)^2}{1 - e^{-2jkh} + j\alpha} \tag{6.85}$$

Equivalent Circuit Parameters

Let us compare the low-frequency equivalent parameters (6.83) with the local quasistatic approximation [27]. The value of inductance L is the same as in [27], but the equivalent capacitance C is different, since it takes into account cell interactions. The local and quasistatic estimation of C from [27, p. 39] is as follows:

$$C = \frac{D\epsilon_0(\epsilon_r + 1)}{\pi} \log\left[\left(\frac{2D}{w}\right) + \sqrt{\left(\frac{2D}{w}\right)^2 - 1}\right] \approx \frac{D\epsilon_0(\epsilon_r + 1)}{\pi} \log\frac{4D}{w} \tag{6.86}$$

Note that (6.86) in [27] was derived from the electric flux density per unit length in a gap between two metallic half-planes. This capacitance, being infinite, urges us to truncate the electric field flux at distance D from the slit to obtain the capacitance per unit length between two coplanar strips of width D. However, the capacitance per unit length of two coplanar strips of width D can be found exactly using the conformal mapping method. For the case $w \ll D$, its expression via the elliptic integrals can be simplified to

$$C = \frac{2D\epsilon_0(\epsilon_r + 1)}{\pi} \log\frac{4}{\sqrt{1 - k'^2}} \tag{6.87}$$

where $k' = \frac{1}{1+w/D}$. For very small w/D, this reduces to

$$C \approx \frac{D\epsilon_0(\epsilon_r + 1)}{\pi} \log\frac{8D}{w} \tag{6.88}$$

As compared to a more accurate expression (6.83), correction can be quite essential, depending on the geometry.

6.4.2   Generalizations

Equation (6.81) has been derived for small periods of the patch array, compared to the wavelength. Also, the period cannot be be large compared to the distance to the ground. More general models will be discussed next.

Sparse Arrays

Restriction $D \ll \lambda$ can be easily lifted for the case when the system is in free space ($\epsilon_r = 1$), since an accurate expression for the grid parameter α is available (see [35] and Chapter 4):

$$\alpha = \frac{kD}{\pi} \left[\log \frac{2D}{\pi w} + \frac{1}{2} \sum_{n=-\infty}^{\infty}{}' \left(\frac{2\pi}{\sqrt{(2\pi n)^2 - k^2 D^2}} - \frac{1}{|n|} \right) \right] \qquad (6.89)$$

(the term with $n = 0$ is excluded from the summation). Substitution of (6.89) into (6.81) defines the array equivalent impedance without restriction on the patch size. Note that for $D > \lambda$ the surface impedance has a nonzero real part, which describes diffraction loss due to excitation of grating lobes. For moderate kD, we can use the Taylor expansion of the series in (6.89):

$$\alpha = \frac{kD}{\pi} \left[\log \frac{2D}{\pi w} + \frac{\zeta(3)}{2} \left(\frac{kD}{2\pi} \right)^2 + \frac{3\,\zeta(5)}{8} \left(\frac{kD}{2\pi} \right)^4 + \cdots \right] \qquad (6.90)$$

where $\zeta(x)$ is the Riemann zeta function.

Higher-Order Floquet Modes Influence

It is known that the field scattered by dense wire and strip meshes excited by plane waves does not practically differ from a plane wave at distances considerably larger than the grid period D (Chapter 4). It means that even if h is small compared to the wavelength but large compared to D, we can consider the interaction between the grid and the metal plane as the far-zone one. It leads to the transmission-line formula for the equivalent surface impedance of the grid parallel to a metal plane, which was used above and in [30]. However, if h becomes smaller than D, we should take into account the influence of higher-order (evanescent) Floquet modes reflected by the ground plane.

    Let us again consider a single planar grid of thin parallel wires or narrow strips in free space excited by a normally incident plane wave polarized along the wires. The grid period D is assumed to be small compared to λ, and the diameter of wires (or the width of strips) is small compared to D. The field scattered by the grid does not essentially differ from that of a square mesh since the orthogonal array of thin wires is practically not excited. The scattered field can be calculated exactly (for thin strips) using the Poisson summation rule (4.11):

$$E_g(y,z) = -\frac{k\eta}{4}I \sum_{n=-\infty}^{\infty} H_0^{(2)}\left(k\sqrt{z^2 + (y-nD)^2}\right)$$

$$= -\frac{k\eta}{2}\frac{I}{D} \sum_{m=-\infty}^{\infty} \frac{e^{-j\frac{2\pi y}{D}m - j\sqrt{k^2 - \left(\frac{2\pi m}{D}\right)^2}|z|}}{\sqrt{k^2 - \left(\frac{2\pi m}{D}\right)^2}} \qquad (6.91)$$

Here, I is the induced current in strips related with the averaged surface current $\widehat{J}$ as $I/D = \widehat{J}$; z and y are coordinates of the observation point with respect to the reference wire positioned at the origin $y = z = 0$ (axis z is orthogonal to the grid plane). For dense grids ($D \ll \lambda$), we can approximate $\sqrt{k^2 - \left(\frac{2\pi m}{D}\right)^2} \approx -j\frac{2\pi m}{D}$ for $m \neq 0$; then the previous formula simplifies, because the series can be expressed in closed form:

$$E_g(y,z) = -\frac{k\eta}{2}\frac{I}{D}\left[\frac{e^{-jk|z|}}{k} + jS(y,z)\right] \qquad (6.92)$$

with

$$S = \frac{D}{\pi}\mathrm{Re}\left\{\sum_{m=1}^{\infty} \frac{e^{-j\frac{2\pi y}{D}m}e^{-\frac{2\pi|z|}{D}m}}{m}\right\} = -\frac{D}{\pi}\mathrm{Re}\left\{\log\left(1 - e^{-\frac{2\pi}{D}(jy+|z|)}\right)\right\} \qquad (6.93)$$

The first term in the expression for E_g is the uniform part of the the scattered field (reflected and transmitted plane waves), and the second term proportional to S is its fluctuating part containing all the evanescent modes.[8]

Next, we consider the same grid of wires at height h over a metal plane. To find the total scattered field, we apply the image principle. The influence of the ground plane is represented as the field reradiated by the image grid plus the image of the incident wave. The total field at the surface of the reference wire is

$$E_{\mathrm{tot}} = -\frac{k\eta}{2D}\left[\frac{e^{-jkr_0}}{k} + jS(0,r_0)\right]I$$

$$+ \frac{k\eta}{2D}\left[\frac{e^{-2jkh}}{k} + jS(0,2h)\right]I + E_{\mathrm{inc}}(1 - e^{-2jkh}) \qquad (6.94)$$

where E_{inc} is the incident electric field. If the wire radius r_0 or the equivalent strip width $w = 4r_0$ is small compared to h, we have $S(0,r_0) \approx \frac{D}{\pi}\log\frac{D}{2\pi r_0}$.

---

[8]This approach leading to the derivation of averaged boundary conditions for wire grids was introduced by A.A. Sochava in his diploma thesis [36].

The boundary condition on conducting wires $E_{\text{tot}} = 0$ allows us to find the induced current I. Finally, writing the reflected field as

$$E_{\text{ref}} = -\frac{\eta I}{2D}\left(1 - e^{-2jkh}\right) - E_{\text{inc}}e^{-2jkh} \tag{6.95}$$

we find that (6.85) is replaced by

$$R = -e^{-2jkh} - \frac{\left(1 - e^{-2jkh}\right)^2}{1 - e^{-2jkh} + j(\alpha + \gamma)} \tag{6.96}$$

with

$$\gamma = \frac{kD}{\pi}\log\left(1 - e^{-\frac{4\pi h}{D}}\right) < 0 \tag{6.97}$$

The only influence of the evanescent modes is the substitution $(\alpha + \gamma)$ instead of α. For the case when the inequality $D \ll \lambda$ is not valid, the correction term cannot be calculated in closed form, but an expression via a very quickly convergent series is available [37]:

$$\gamma = \frac{kD}{\pi}\sum_{m=1}^{\infty}\left(\frac{1 - e^{-4\pi\frac{h}{D}\sqrt{m^2 - \left(\frac{kD}{2\pi}\right)^2}}}{\sqrt{m^2 - \left(\frac{kD}{2\pi}\right)^2}} - \frac{1}{m}\right) \tag{6.98}$$

The higher-order modes influence turns out to be negligible even if $h \sim D/2$, and it becomes essential only if the thickness h is small compared to D.

This theory is based on an estimation for the grid impedance of arrays of thin conducting strips (6.73) and (6.74). Although the theory takes into account dynamic interaction of the strips in the infinite arrays, the local field distribution near a strip is assumed to be quasistatic.[9] Under this approximation the grid impedance (6.73) is purely inductive and its value does not depend on the permittivity of the medium in which the grid is located. The grid impedance of the complementary grid of patches (6.77) is, naturally, a pure capacitance that does not feel the permeability of the medium. A more accurate theory of artificial impedance layers can be built using the exact solution for the field of an infinitely long perfectly conducting wire (or a thin strip) near an interface. The results show that for the case of moderate contrasts ($\epsilon_r < 3$), the correction is very small [31].

6.4.3 Comparison with the Local Quasistatic Model and Experiment

An artificial impedance surface with square patches was studied in [26] using numerical techniques. In particular, dispersion curves for surface waves along

---

[9]The nature of the approximation can be seen from the derivation of (6.96). Repeating the derivation in the absence of the ground plane, we arrive at (6.72) and (6.74).

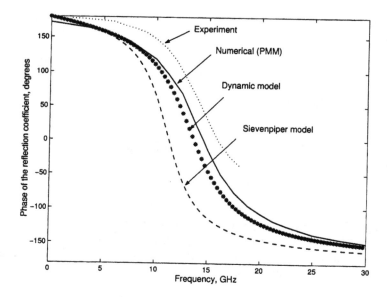

Figure 6.17 Frequency behavior of the phase of the reflection coefficient. Dotted and solid lines show numerical and experimental results of paper [38]. Dashed line has been calculated using the quasistatic formula (6.86) introduced in [26].

the structure were plotted in Figure 10 of that paper. In that figure, the point at which the TE surface wave curve crosses the speed-of-light line corresponds to the resonant frequency of the equivalent surface impedance.[10] For the example considered in [26], numerical simulations give the resonant frequency at approximately 14.4 GHz, the present model gives 14.37 GHz, and the local and quasistatic model [27] gives 11.65 GHz.

A more detailed comparison can be made with data for the phase of the reflection coefficient at normal incidence, which is available in [38]. In that work, the reflection coefficient was calculated numerically using the periodical method of moments and measured experimentally. The surface parameters are the following: $D = 2.44$ mm, $w = 0.15$ mm, $\epsilon_r = 2.51$, and $h = 1.57$ mm. We have extracted the numerical and experimental data from the corresponding plots in [38, Figure 32a, page 67, and Figure 36a, page 81] and plotted them together with the data calculated using the dynamic analytical model (Figure 6.17). For comparison, the corresponding results of the local quasistatic model [26] are also shown. Obviously, the dynamic model more closely matches the numerical and experimental data.

---

[10]TE surface waves are guided by the impedance surface at higher frequencies; see Section 6.4.5.

6.4.4   Oblique Incidence on Mushroom Structures

Let us study the equivalent surface impedance and the reflection properties of high-impedance surfaces in the form of mushroom layers for obliquely incident plane waves. Here we should distinguish between two polarizations of the incident fields, considering plane waves propagating along one of the crystal axes (Figure 6.18).

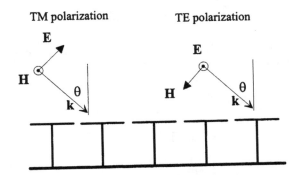

Figure 6.18 TM and TE plane waves incident on a mushroom layer.

The grid impedance modeling patch arrays for oblique incidence can be found from the grid impedance for strip arrays and the Babinet principle. Inspecting (4.38) for the averaged boundary condition for wire or strip gratings, we see that for TE polarized fields

$$Z_g^{TE} = j\frac{\eta}{2}\alpha \tag{6.99}$$

[angle $\varphi = \pi/2$ in (4.38)], and for TM polarized fields

$$Z_g^{TM} = j\frac{\eta}{2}\alpha \cos^2\theta \tag{6.100}$$

(angle $\varphi = 0$). Applying the Babinet principle (4.59), we must remember that in this transformation electric and magnetic fields replace each other. Thus, to find the grid impedance for patch arrays in TE-polarized fields, we should use the Babinet transformation for strip arrays in TM fields. This way we find for patch arrays

$$Z_g^{TE} = \frac{\eta}{2j\alpha\cos^2\theta} \tag{6.101}$$

for TE polarized excitation, and

$$Z_g^{TM} = \frac{\eta}{2j\alpha} \tag{6.102}$$

for TM waves. Of course, in the formula for the grid parameter α the strip width should be replaced by the gap width; compare (4.58) and (4.60).

TE polarized incidence is easier to analyze because in this case there is no current in vias connections, and we can consider the space between the patches and the ground as an isotropic dielectric slab. The equivalent surface impedance at the position of the patch array is given by (2.109):

$$Z_s^{TE} = j\omega\mu\frac{\tan\beta h}{\beta} \approx j\omega\mu h \tag{6.103}$$

for thin slabs with $|\beta|h = |k|h\cos\theta \ll 1$. The input impedance of the whole structure is the parallel connection of Z_g^{TE} and Z_s^{TE}:

$$Z_{\text{inp}}^{TE} = \frac{jkh\eta}{1 - 2kh\alpha\cos^2\theta} \tag{6.104}$$

Because the wave impedance for TE waves is $\eta/\cos\theta$ (2.17), the reflection coefficient reads

$$R^{TE} = \frac{Z_{\text{inp}}^{TE}\cos\theta - \eta}{Z_{\text{inp}}^{TE}\cos\theta + \eta} \tag{6.105}$$

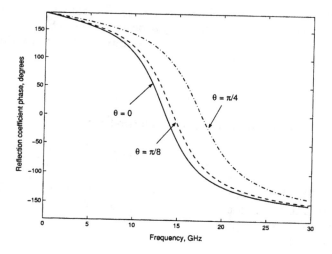

Figure 6.19 Reflection coefficient phase for different incidence angle of TE-polarized incident plane waves. The geometry and dimensions of the structure are the same as in Figure 6.17.

Figure 6.19 shows how the reflection coefficient phase changes with the incidence angle for TE-polarized fields. For oblique incidence the resonance frequency becomes higher, and for grazing angles the surface looks like metal.

For TM incident fields the analysis is slightly more involved, because the effect of vertical pins is important.[11] The array of thin conducting pins between the patch array and the ground can be viewed as a slab filled by a wire medium.[12] In this medium, two eigenwave solutions exist. One is the TM mode of the wire medium, which is exponentially decaying because the operational frequency is below the equivalent plasma frequency and the effective permittivity is negative (see Section 5.5.4). The influence of this mode can be neglected in this case. Indeed, its influence is manifested by some charge concentration near the wire ends, which is not important because the additional charge on thin wires is very small compared with the charge accumulated on large metal patches. Thus, the only important solution in the space between the patch array and the ground is the TEM wave. Its propagation factor has two components. One is orthogonal to the patch array plane and it is equal to the wavenumber in the substrate dielectric. The other is tangential to that plane and naturally equal to the tangential component of the wavenumber of the incident field. Electric and magnetic fields of this mode are orthogonal to the wires and their ratio (wave impedance) for the case of thin pins is approximately the same as in the filling dielectric. This is because the electric polarization of thin wires in the transverse direction is small, and the averaged quasistatic magnetic field of the wire currents is zero due to the symmetry of the problem. Thus, we conclude that the surface impedance seen at the input plane of the wire grid is simply the same as that of a TEM transmission line section of the length equal to the layer thickness. The influence of pins is seen in the fact that this surface impedance does not depend on the incidence angle:

$$Z_s^{TM} = Z_s^{TE} = j\omega\mu\frac{\tan\beta h}{\beta} \approx j\omega\mu h \qquad (6.106)$$

Note here that the equivalent surface impedance of a dielectric slab on a metal surface does depend on the incidence angle as $\cos^2\theta$, if the wave is TM-polarized, see (2.109). The input impedance of the whole structure is the parallel connection of this surface impedance and the grid impedance of the patch array Z_s^{TM} (6.102):

$$Z_{\text{inp}}^{TM} = \frac{jkh\eta}{1 - 2kh\alpha} \qquad (6.107)$$

The resonance frequency of this surface impedance does not depend on the incidence angle; compare with (6.104). Finally, the reflection coefficient is

$$R = \frac{Z_{\text{inp}}^{TM} - \eta\cos\theta}{Z_{\text{inp}}^{TM} + \eta\cos\theta} \qquad (6.108)$$

---

[11]In the absence of vias connectors, the TM solution for the reflection coefficient is exactly the same as for TE waves for all incidence angles.

[12]This model has been developed by S.I. Maslovski.

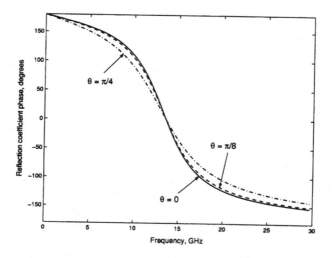

Figure 6.20 Reflection coefficient phase for different incidence angles of TM-polarized incident plane waves. TEM-mode model.

Figure 6.20 shows how the reflection coefficient phase depends on the frequency for different incidence angles for TM-polarized fields. The effect of pins is clearly seen in much more stable properties with respect to the incidence angle as compared to the case when there are no vias connectors.

An alternative model can be developed using the fact that in practical situations the distance between vias connectors is much smaller than the wavelength, so it is possible to use an effective medium model to replace this complicated wire structure (this approach was suggested in [30]). We know that wire arrays with long wires possess strong spatial dispersion. However, in this particular system the wires are rather short (much shorter than the wavelength), so the quasistatic model of wire media of Section 5.5.4 is more appropriate. The boundary condition for a slab of wire medium on an ideally conducting surface can be derived in the same way as for isotropic slabs (Problem 2.3). The result is

$$\mathbf{E}_{t+} = j\omega\mu \frac{\tan(\beta_{TM}h)}{\beta_{TM}} \left(1 - \frac{k_t^2}{\omega^2\epsilon_n\mu}\right) \mathbf{n} \times \mathbf{H}_{t+} \qquad (6.109)$$

where (2.20)

$$\beta_{TM}^2 = \omega^2\epsilon_t\mu - \frac{\epsilon_t}{\epsilon_n}k_t^2 \qquad (6.110)$$

and $k_t^2 = k^2 \sin^2\theta$. For thin substrates with $|\beta_{TM}|h \ll 1$, we have approximately

$$Z_s^{TM} = jkh\eta \left(1 - \frac{k_t^2}{\omega^2\epsilon_n\mu}\right) \qquad (6.111)$$

The transverse component of the effective permittivity dyadic equals that of the matrix dielectric if the pins are thin, and the normal component is given by the plasma-like formula (5.210). Substituting the permittivity ϵ_n, we get

$$Z_s^{TM} = jkh\eta \frac{k^2 \cos^2 \theta - k_p^2}{k^2 - k_p^2} \qquad (6.112)$$

For low frequencies we have $k \ll k_p$, and the result is close to that of the transmission-line model (6.106).

Finally, the input impedance and the reflection coefficients are found in the same way as for TE waves. Calculations give curves very similar to that in Figure 6.20, although in this model the resonance frequency slightly changes with changing incidence angle. For thicker substrates, (6.106) is more accurate.

6.4.5   Surface Waves Along Impedance Surfaces

For many applications the properties of surface waves along layers are very important. Most often, excitation of surface waves is not desirable. Surface waves along substrates of microstrip antennas distort the radiation pattern and reduce antenna efficiency. Surface waves in substrates of microwave integrated circuits increase cross talk. Here we will study surface waves along impedance surfaces.

Consider surface waves along a planar interface with a given isotropic surface impedance Z_s (Figure 6.21). That is, on the surface $x = 0$ the following boundary condition is satisfied:

$$\mathbf{E}_t = Z_s \, \mathbf{x}_0 \times \mathbf{H} \qquad (6.113)$$

In Cartesian components, we have

$$E_z = Z_s H_y \quad \text{TM waves,} \qquad E_y = -Z_s H_z \quad \text{TE waves} \qquad (6.114)$$

We assume for simplicity that the half-space $x > 0$ is filled by air, with the parameters ϵ_0 and μ_0.

Let us look for solutions in form of a surface wave, that is, a wave that exponentially decays with increasing the distance from the surface x (fields are zero behind the *boundary*, at $x < 0$). Starting from TE waves, we write

$$E_y = A e^{-j\beta z - \alpha x} \qquad (6.115)$$

(no field dependence along y — we can always direct the y-axis orthogonally to the wave vector). Since every field component satisfies the Helmholtz equation

$$(\nabla^2 + k_0^2) E_y = 0 \qquad (6.116)$$

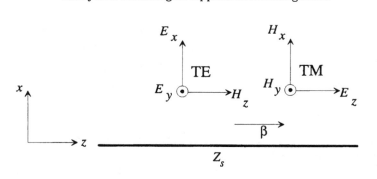

Figure 6.21 Geometry of the problem: surface TE and TM waves along a planar impedance surface.

we have

$$\alpha^2 - \beta^2 + k_0^2 = 0 \tag{6.117}$$

Magnetic field components follow from the Maxwell equation:

$$\mathbf{H} = -\frac{1}{j\omega\mu_0}\nabla \times \mathbf{E} \tag{6.118}$$

that gives

$$H_z = \frac{\alpha}{j\omega\mu_0}E_y \tag{6.119}$$

Applying the boundary condition (6.114), we find the decay factor

$$\alpha = -\frac{j\omega\mu_0}{Z_s} \qquad \text{TE waves} \tag{6.120}$$

Thus, the propagation factor is, from (6.117),

$$\beta = k_0^2 - \frac{\omega^2\mu_0^2}{Z_s^2} = k_0^2\left(1 - \frac{\eta_0^2}{Z_s^2}\right) \qquad \text{TE waves} \tag{6.121}$$

where $k_0 = \omega\sqrt{\epsilon_0\mu_0}$ and $\eta_0 = \sqrt{\mu_0/\epsilon_0}$.

In the same way, for TM waves we find that

$$\alpha = -j\omega\epsilon_0 Z_s \qquad \text{TM waves} \tag{6.122}$$

$$\beta = k_0^2\left(1 - \frac{Z_s^2}{\eta_0^2}\right) \qquad \text{TM waves} \tag{6.123}$$

Propagating surface waves have purely real propagation constants β, as defined in (6.115). Naturally, it is only possible if the surface impedance is purely imaginary; that is, the surface is lossless, see (6.121) and (6.123). In this case, the decay factor α is purely real [see (6.120) and (6.122)]. Moreover, only surfaces whose impedance has a positive imaginary part (e.g., inductive impedance $Z_s = j\omega L$) can support TM surface waves and only surfaces with negative reactance can support TE waves. This is obvious from (6.120) and (6.122); for example, if $Z_s = j\omega L$, decay factor for TE waves becomes negative, which means radiation (leaky waves). Clearly, the value of the impedance determines how much the field is confined to the surface. One of the applications of artificial impedance surfaces is a wire antenna parallel to the surface. Such an antenna directed along y can radiate only TE waves. As TE surface waves are not supported by an *inductive* surface, the antenna works properly if the surface is inductive.

Let us now consider the dispersion relation for surface waves over an artificial impedance surface. Frequency dependence of the surface impedance was considered previously. Approximately, it has the resonant form:

$$Z_s = \frac{j\omega L}{1 - \omega^2 LC} = \frac{j\omega L}{1 - \left(\frac{\omega}{\omega_0}\right)^2} \tag{6.124}$$

Using (6.121) and (6.123), we can visualize the frequency behavior of the propagation factors as functions of the frequency (see Figure 6.22). Practically, near the resonance frequency there is a stopband for surface waves (in the region where the propagation factor for the TM wave is very large but TE waves still cannot propagate).

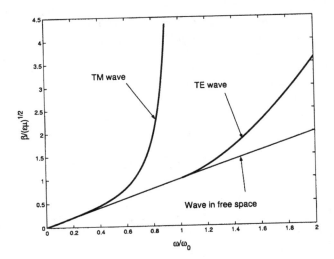

Figure 6.22 Dispersion curves for TE and TM surface waves over an artificial impedance surface.

Consider now waves along highly conducting surfaces. In this case, using the Leontovich impedance boundary condition (3.7), we have

$$Z_s = \sqrt{\frac{\mu}{\epsilon}} \approx \sqrt{\frac{j\omega\mu}{\sigma}} \qquad (6.125)$$

where ϵ and μ are the parameters of the conductor, σ is its conductivity, and we have neglected ϵ as compared with σ/ω. The surface impedance is inductive, and TM waves can be supported.

The propagation factor is of course complex because the surface is lossy:

$$\beta^2 = k_0^2 \left(1 - j\epsilon\frac{\omega}{\sigma}\right) \qquad (6.126)$$

For good conductors (even for the Earth at radio frequencies), $\frac{\sigma}{\omega} \gg \epsilon$; thus, $\beta \approx k_0$. But this means that $\alpha^2 = \beta^2 - k_0^2 \approx 0$ [we use (6.117)] and the wave is not confined to the surface. Actually, this is just a usual plane wave along the surface.

6.4.6 Antenna Applications of Artificial Impedance Surfaces

Introducing artificial high-impedance surfaces, we mentioned that they can be used in antennas for size reduction of small antennas and for screening of mobile phone radiation from the user's head [26]. The first question to ask is: What impedance value is the best suited for this application [29]? To study the problem we make use of the infinite-plane model of the structure. Considering the field distribution in the near zone of a wire antenna, we can make conclusions regarding the possible size of the screen. Indeed, since the induced current is proportional to the magnetic field values on the impedance screen, the areas where the electromagnetic field is small are not essential for the operation of the device.

Consider an infinite line current positioned at distance h from an impedance surface (Figure 6.23) modeled by an isotropic impedance Z_s:

$$\mathbf{E}_t = Z_s\mathbf{n} \times \mathbf{H} = \eta\overline{Z}_s\mathbf{n} \times \mathbf{H} = \frac{\eta}{\overline{Y}_s}\mathbf{n} \times \mathbf{H} \qquad (6.127)$$

where $\overline{Z}_s = Z_s/\eta$ and $\overline{Y}_s = Y_s\eta$ are the surface impedance and admittance, respectively, that are normalized to the free-space impedance $\eta = \sqrt{\mu_0/\epsilon_0}$.

To calculate the electromagnetic field we make use of the exact image theory [15]. The image current [15, Eq. (7.165)]

$$I_i = \delta_+(\zeta)I - 2jk\overline{Y}_se^{-j\overline{Y}_sk\zeta}U_+(\zeta)I \qquad (6.128)$$

contains two parts: a line current in the geometrical image position and a distributed image current that can be considered as located in complex space.

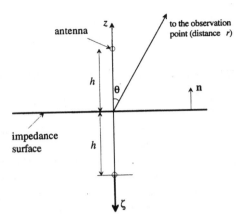

Figure 6.23 Geometry of the problem.

The image current is concentrated on the axis ζ that originates from the image point and extends to minus infinity in the vertical direction (Figure 6.23). In (6.128), δ_+ is the Dirac delta function and U_+ is the unit step function. Obviously, for the ideal magnetic wall ($Y_s = 0$) the distributed image current vanishes.

The line current I creates electric field (directed along the current) [39, Sec. 5.4]

$$E = -\frac{\eta k}{4} H_0^{(2)}(kR)I \qquad (6.129)$$

In the far zone,

$$E = -\frac{\eta k}{\sqrt{8\pi kR}} e^{-j(kR-\pi/4)} I \qquad (6.130)$$

Let us calculate the field radiated in the far zone by the line current over the impedance surface. Denoting the distance from the origin to the observation point by r, we approximate the distance from the original current to the observation point as

$$R = r - h\cos\theta \qquad (6.131)$$

From the geometrical image point,

$$R = r + h\cos\theta \qquad (6.132)$$

From an arbitrary point at the line $0 < \zeta < \infty$,

$$R = r + (h + \zeta)\cos\theta \qquad (6.133)$$

The total radiated field in the far zone is created by the original current line, the image line at the symmetric position, and the distributed image current:

$$E = -\frac{\eta k}{\sqrt{8\pi}} e^{j\pi/4} \left\{ \frac{e^{-jk(r-h\cos\theta)}}{\sqrt{k(r-h\cos\theta)}} + \frac{e^{-jk(r+h\cos\theta)}}{\sqrt{k(r+h\cos\theta)}} \right.$$

$$-2jk\overline{Y_s}\left.\int_0^\infty \frac{e^{-jk[\overline{Y_s}\zeta+r+(h+\zeta)\cos\theta]}}{\sqrt{k[r+(h+\zeta)\cos\theta]}}d\zeta\right\}I \tag{6.134}$$

Assuming lossless upper half-space, the integral converges if

$$\text{Im}\{\overline{Y_s}\} < 0 \qquad \text{or} \qquad \text{Im}\{\overline{Z_s}\} > 0 \tag{6.135}$$

(positive reactance, e.g., inductance). This physically means that there are no surface waves excited on the impedance surface. Indeed, we know from Section 6.4.5 that no TE waves are supported if (6.135) is satisfied. In our particular case only TE modes can exist. The case when surface modes do not exist is of practical interest, so we concentrate on that situation.

To calculate the far-field pattern, relation (6.134) can be further simplified, because for large kr we can approximate the denominator of the integrand in (6.134) by $\sqrt{kr}$. The result is, for the antenna pattern,

$$F(\theta) = \left|1 + e^{-2jkh\cos\theta} - \frac{2\overline{Y_s}}{\overline{Y_s}+\cos\theta}e^{-2jkh\cos\theta}\right| \tag{6.136}$$

or

$$F(\theta) = \left|1 + \frac{\cos\theta - \overline{Y_s}}{\cos\theta + \overline{Y_s}}e^{-2jkh\cos\theta}\right| \tag{6.137}$$

Here, the pattern is normalized to the field of the single wire in free space bearing current I.

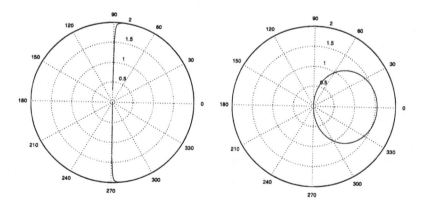

Figure 6.24 Field pattern of a line source near a high-impedance surface. Left: a high value of inductive surface impedance $\overline{Z_s} = 100j$; right: a low level of inductive surface impedance $\overline{Z_s} = j$.

The pattern for $kh = 0.1$ and $\overline{Z_s} = j100$ is plotted in Figure 6.24, left. Obviously, for this case the surface acts nearly as a magnetic wall, so the radiated field is about doubled due to the reflection in the wall. The effect in the case of low values of the surface admittance can be seen from Figure 6.24, right. Here, $\overline{Z}_s = j$. Also in this case the radiation is enhanced in the direction orthogonal to the surface, although the impedance value is quite small.

Because in practice we have a limited size of the screen (and usually it is desirable to make the screen size as small as possible), we should consider the field pattern in the near zone as well. Indeed, it is necessary to cut the reflection screen in the near zone, since its size practically cannot be made large compared to the wavelength. It is well known that the pattern of the far field and that of the near field can dramatically differ.

For an arbitrary point in space, defined by the coordinates r, θ (Figure 6.23), the total field is

$$E = -\frac{\eta k}{4} \left[H_0^{(2)} \left(k\sqrt{r^2 - 2rh\cos\theta + h^2} \right) \right.$$

$$\left. + H_0^{(2)} \left(k\sqrt{r^2 + 2rh\cos\theta + h^2} \right) - 2j\overline{Y}_s G \right] I \qquad (6.138)$$

where

$$G = \int_0^\infty e^{-j\overline{Y}_s \zeta} H_0^{(2)} \left(\sqrt{(kr\cos\theta + kh + \zeta)^2 + (kr\sin\theta)^2} \right) d\zeta \qquad (6.139)$$

The field pattern, normalized to the field of a single wire in free space, reads

$$F(kr, \theta) = 1 + \frac{H_0^{(2)} \left(\sqrt{(kr)^2 + 2k^2 rh\cos\theta + (kh)^2} \right) - 2j\overline{Y}_s G}{H_0^{(2)} \left(\sqrt{(kr)^2 - 2k^2 rh\cos\theta + (kh)^2} \right)} \qquad (6.140)$$

Because of the exponential factor, the integral in (6.139) converges quickly, and can be calculated numerically.

In Figures 6.25 and 6.26, the transformation of the field pattern in the near zone is shown. The near-zone concentration of the field in the desired direction very much depends on the surface impedance value. Comparing these two figures, we note that for smaller values of the imaginary part of the surface impedance we have a better shape on the field pattern in the near zone (which means that we can have a smaller screen size), but the trade-off is the directivity in the far zone, which becomes smaller in this case. Essentially, this means that there exists an optimum of the surface impedance corresponding to a compromise between higher levels of the antenna gain and radiation resistance and lower levels of the field at the screen plane (which means smaller screen sizes).

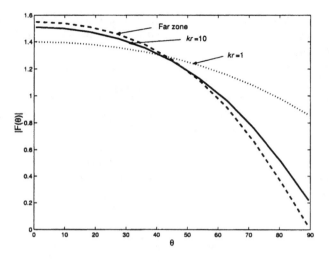

Figure 6.25 Field pattern of a line source near a low-impedance surface. Near-field pattern for $\overline{Z_s} = j$.

6.5   MATERIALS WITH NEGATIVE PERMITTIVITY AND PERMEABILITY

V.G. Veselago studied in the 1960s the electromagnetic properties of materials whose permittivity and permeability simultaneously have negative real parts and showed that these media exhibit unusual properties, such as support of backward waves (when the Poynting vector of a time-harmonic plane wave is directed opposite with respect to the wave vector, Section 5.3.2) and anomalous negative refraction of plane monochromatic electromagnetic waves [40]. This concept implies many other interesting features of wave propagation that were reviewed in [40]. Most new effects are the consequences of the fact that this is a backward-wave material. We know, however, that backward waves can also exist for example in periodical structures (Chapter 5). The fundamental difference between the *Veselago medium* and other backward-wave structures is that here we deal with *homogeneous materials,* whose microstructure can be averaged on the scale of the wavelength and whose properties can be described by effective permittivity and permeability.

It appears that Veselago media do not occur in nature: Earlier attempts to find realizations among magnetic semiconductors failed mainly due to high losses in those materials. The use of the concept of metamaterials opens a way to realize such exotic materials. Since it is easier to use resonant inclusions with millimeter-scale dimensions, it seems reasonable to focus on the realizations of Veselago media for the microwave region of the electromagnetic spectrum (optical composites will need very small particles) and

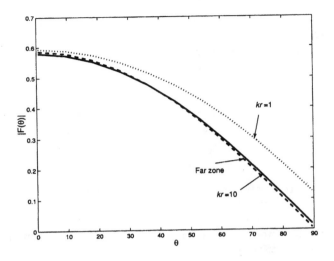

Figure 6.26 Field pattern of a line source near a low-impedance surface. Near-field pattern for $\overline{Z}_s = 0.2j$.

to study their response experimentally. This work started at the University of California at San Diego and the first positive experimental results were published in [41] (transmission through slabs) and [42] (negative refraction). These experiments have been based on numerical simulations and a simple analytical model presented in [41, 43].

6.5.1 Basic Electromagnetic Properties of Veselago Media

The main property that is specific to media with negative material parameters is that these media support plane backward waves. For a homogeneous plane wave in a lossless material $e^{j\omega t - \mathbf{k}\cdot\mathbf{r}}$ with a real vector $\mathbf{k}$ in an isotropic medium, the Maxwell equations read

$$\mathbf{k} \times \mathbf{E} = \omega\mu\mathbf{H}, \qquad \mathbf{k} \times \mathbf{H} = -\omega\epsilon\mathbf{E} \tag{6.141}$$

Vector multiplying the first equation by $\mathbf{E}$ and the second one by $\mathbf{H}$, we find the Poynting vector of the wave:

$$\mathbf{S} = \mathbf{E} \times \mathbf{H} = \frac{E^2}{\omega\mu}\mathbf{k} = \frac{H^2}{\omega\epsilon}\mathbf{k} \tag{6.142}$$

In conventional isotropic lossless media with real and *positive* material parameters the directions of the Poynting vector and the wave vector $\mathbf{k}$ are of course the same. But if the material parameters are negative, these two vectors have the opposite directions. This is illustrated in Figure 6.27. Three vectors **E**, **H**, and **k** now form a left-handed system, which is why the materials with negative parameters are sometimes called left-handed media.

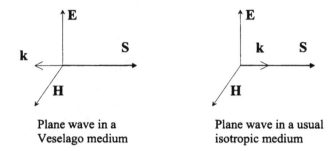

Plane wave in a   Plane wave in a usual
Veselago medium   isotropic medium

Figure 6.27 Homogeneous plane waves in isotropic media. If the material parameters are negative, the wave is a *backward wave*.

Unusual properties can be seen at reflection from an interface between a usual and a backward-wave medium (see Figure 6.28). The law of refraction follows from two requirements. First, the tangential component of the wavevector **k** must be the same in two media, so that the boundary conditions be satisfied. Second, the Poynting vector in the second medium must be directed away from the interface, because the source is in the first medium. From here we conclude that the plane wave is refracted "negatively" or "anomalously" if one of the two media supports backward waves.

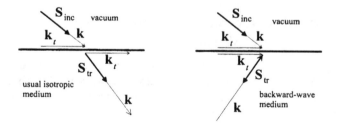

Figure 6.28 Refraction of plane monochromatic waves at interfaces between two isotropic media. "Negative refraction" occurs if one of the two media is a backward-wave medium.

There have been contradictory opinions on the subject of negative refraction in the literature [44]. The confusion originates from the fact that the directions of the power flow and the signal front (in case of space-time modulated signals) are different.[13] Figure 6.29 (drawn by S.I. Maslovski) schematically shows how a space-time modulated pulse of a high-frequency

---

[13]The following clarification belongs to S.I. Maslovski.

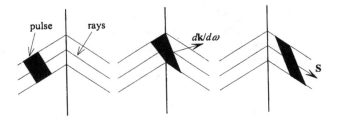

Figure 6.29 Propagation of a space-time modulated pulse. The images show the pulse at increasing moments of time, from left to right.

signal interacts with a backward-wave medium interface. The pulse is represented by a black rectangle. We can assume that the blackened area represents most of the pulse energy, and that the pulse itself is smooth enough to have most of the spectrum energy in the backward-wave frequency region. The signal fronts propagate causally [44], which means that for these fronts a law similar to the Huygens principle is applicable. At the same time the rays refract negatively, since the total energy goes this way. These two simple facts explain the phenomenon: From the figure we see that the pulse forward and backward fronts go along the time delay vector $d\mathbf{k}/d\omega$, where $\mathbf{k}$ is the wave vector, *but* the entire pulse still propagates in the direction of the energy flow, according to the negative refraction law (see more in [45–47]).

Because in any passive medium the energy density function must be nonnegative definite, any passive Veselago medium should be dispersive and must satisfy constraints (e.g., [48])

$$\frac{d[\omega\epsilon(\omega)]}{d\omega} > 1, \qquad \frac{d[\omega\mu(\omega)]}{d\omega} > 1 \qquad (6.143)$$

These constraints are valid in the frequency regions where dissipation is small, so that the energy density can be defined.

6.5.2 Slabs of Veselago Media: Impedance Boundary Conditions

Properties of thin layers of materials with negative parameters can be analyzed using the results of Section 2.2 without any modification, since the derivation was general and we made no assumption about material parameter values. Thus, modeling thin slabs we can use (2.34) and (2.35) for a slab of thickness d

$$\mathbf{E}_{t+} - \mathbf{E}_{t-} = \left(j\omega\mu d \overline{\overline{I}}_t - \frac{d}{j\omega\epsilon}\nabla_t\nabla_t \right) \cdot \mathbf{n} \times \mathbf{H}_{t-} \qquad (6.144)$$

$$\mathbf{n} \times \mathbf{H}_{t+} - \mathbf{n} \times \mathbf{H}_{t-} = j\omega\epsilon d\, \mathbf{E}_{t-} + \frac{d}{j\omega\mu}\nabla_t \times (\nabla_t \times \mathbf{E}_{t-}) \qquad (6.145)$$

with negative values of ϵ and μ. For negative material parameters the sign of the right-hand side is reversed. In particular, for a thin slab on an ideally conducting surface ($\mathbf{E}_{t-} = 0$) and normally incident plane waves, the boundary condition reads

$$\mathbf{E}_{t+} = j\omega\mu d\, \mathbf{n} \times \mathbf{H}_{t+} \tag{6.146}$$

Note that for thin slabs on conducting surfaces the permittivity value has no importance, which is the same as for usual media. For a slab of a usual magnetodielectric the input impedance is inductive, as with a short section of a shortened transmission line. If the permeability is negative, the input impedance is a *negative* inductance: $Z_s = -j\omega|\mu|d$. This looks very attractive for potential applications in metamaterials: Recall that negative inductances are often needed for the design of broadband structures (negative inductance can compensate normal positive inductance of arrays of conducting wires, for instance). Unfortunately, materials with negative parameters should be themselves created as metamaterials.

The exact solution for an isotropic slab in Section 2.3 is also general and can be used for any values of the material parameters. In the linear relations between tangential fields on the two sides of the slab (2.111)

$$\begin{pmatrix} \mathbf{E}_{t+} \\ \mathbf{n} \times \mathbf{H}_{t+} \end{pmatrix} = \begin{pmatrix} \overline{\overline{a}}_{11} & \overline{\overline{a}}_{12} \\ \overline{\overline{a}}_{21} & \overline{\overline{a}}_{22} \end{pmatrix} \cdot \begin{pmatrix} \mathbf{E}_{t-} \\ \mathbf{n} \times \mathbf{H}_{t-} \end{pmatrix} \tag{6.147}$$

the dyadic coefficients are

$$\overline{\overline{a}}_{11} = \overline{\overline{a}}_{22} = \cos(\beta d)\overline{\overline{I}}_t \tag{6.148}$$

$$\overline{\overline{a}}_{12} = \frac{j\omega\mu}{\beta}\sin(\beta d)\,\overline{\overline{A}}, \qquad \overline{\overline{a}}_{21} = \frac{j\omega\epsilon}{\beta}\sin(\beta d)\,\overline{\overline{C}} \tag{6.149}$$

These are the same equations as given by (2.112) and (2.113); we only replaced $k\eta = \omega\mu$ and $k/\eta = \omega\epsilon$ to stress the explicit dependence on the material parameters. In the exact solution, as in the approximate boundary conditions, reversing the sign of the material parameters leads to reversing the sign of $\overline{\overline{a}}_{12}$ and $\overline{\overline{a}}_{21}$. Because inside the slab there are waves traveling along both directions of the z-axis, the result is independent from the choice of the sign of the normal component of the propagation constant $\beta = \sqrt{k^2 - k_t^2}$.

Example: Two-Layer Waveguides and Resonators

Consider a plane waveguide formed by two slabs of different isotropic materials and bounded by two metal planes (Figure 6.30).

The exact boundary condition on the free interface of a slab backed by an ideally conducting surface reads (2.109):

$$\mathbf{E}_{t+} = j\omega\mu\frac{\tan\beta d}{\beta}\overline{\overline{A}} \cdot \mathbf{n} \times \mathbf{H}_{t+} \tag{6.150}$$

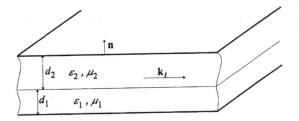

Figure 6.30 Planar two-layer waveguide. One of the slabs can be a Veselago medium.

where

$$\overline{\overline{A}} = \overline{\overline{I}}_t - \frac{\mathbf{k}_t \mathbf{k}_t}{k^2} = \frac{\beta^2}{k^2} \frac{\mathbf{k}_t \mathbf{k}_t}{k_t^2} + \frac{\mathbf{n} \times \mathbf{k}_t \, \mathbf{n} \times \mathbf{k}_t}{k_t^2} \qquad (6.151)$$

This condition can be used to model both slabs. Because the tangential fields $\mathbf{E}_{t+}$ and $\mathbf{n} \times \mathbf{H}_{t+}$ are continuous on the interface between two slabs, we can write

$$\left(j\omega\mu_1 \frac{\tan \beta_1 d_1}{\beta_1} \overline{\overline{A}}_1 + j\omega\mu_2 \frac{\tan \beta_2 d_2}{\beta_2} \overline{\overline{A}}_2 \right) \cdot \mathbf{n} \times \mathbf{H}_{t+} = 0 \qquad (6.152)$$

Here, indices $1, 2$ refer to the two slabs:

$$\overline{\overline{A}}_1 = \overline{\overline{I}}_t - \frac{\mathbf{k}_t \mathbf{k}_t}{k_1^2}, \qquad \overline{\overline{A}}_2 = \overline{\overline{I}}_t - \frac{\mathbf{k}_t \mathbf{k}_t}{k_2^2} \qquad (6.153)$$

and so forth. A solution for the eigenwaves is now very easy because dyadics $\overline{\overline{A}}_{1,2}$ are diagonal with the same set of eigenvectors: $\mathbf{k}_t$ and $\mathbf{n} \times \mathbf{k}_t$. Writing the two-dimensional vector $\mathbf{n} \times \mathbf{H}_{t+}$ in this basis:

$$\mathbf{n} \times \mathbf{H}_{t+} = a \, \mathbf{k}_t / |k_t| + b \, \mathbf{n} \times \mathbf{k}_t / |k_t| \qquad (6.154)$$

and substituting into (6.152), we arrive at equations for the propagation constant k_t. If $a \neq 0$ and $b = 0$, vector $\mathbf{H}_{t+}$ is directed along $\mathbf{n} \times \mathbf{k}_t$; that is, orthogonal to the propagation direction. This gives the TM-mode solution. The eigenvalue equation in this case is

$$\frac{\beta_1}{\epsilon_1} \tan \beta_1 d_1 + \frac{\beta_2}{\epsilon_2} \tan \beta_2 d_2 = 0 \qquad (6.155)$$

For the other mode, when $b \neq 0$ and $a = 0$, the magnetic field vector is along $\mathbf{k}_t$ (TE mode), and we get

$$\frac{\mu_1}{\beta_1} \tan \beta_1 d_1 + \frac{\mu_2}{\beta_2} \tan \beta_2 d_2 = 0 \qquad (6.156)$$

If the propagation constant along the slabs k_t is zero, both equations for TM and TE modes reduce to

$$\frac{\mu_1}{k_1} \tan k_1 d_1 + \frac{\mu_2}{k_2} \tan k_2 d_2 = 0 \qquad (6.157)$$

that is the resonance condition for standing waves in the dual-layer system between two metal plates. If the thicknesses of both layers are small compared with the wavelength, we can simplify this equation replacing tangent functions by the first terms of their Taylor expansions:

$$\mu_1 d_1 + \mu_2 d_2 = 0 \qquad (6.158)$$

From here it is obvious that if both permeabilities are positive (or both negative), no resonance is possible in thin layers: The thickness should be of the order of the wavelength (half-wavelength resonance). However, as was noticed by N. Engheta [49], if *one* of the permeabilities is negative, condition (6.158) can be satisfied even for very thin layers. This appears to open a possibility to realize very compact resonant cavities.

However, what kind of resonator would we get this way? Resonance as such means that a certain circuit function (reflection coefficient, input impedance...) sharply changes with the frequency. In the conventional case of positive media parameters, the two-layer cavity is a *resonator* because the tangent functions in (6.157) quickly vary with respect to the frequency near the resonance. Now look at relation (6.158): There is no explicit dependence on the frequency at all. So, if the material parameters are assumed to be approximately frequency independent over a certain frequency range, the system does not resonate, although the "resonance condition" (6.158) is satisfied. Indeed, (6.158) is satisfied for all frequencies in this range; thus, this system is not frequency selective. What actually happens is that the inductive reactive part of the input impedance of a usual layer $(+j\omega\mu_1 d_1)$ is compensated by the *negative inductive impedance* of the other layer $(-j\omega|\mu_2|d_2)$. What is left is only resistance, defined by losses in the materials of the slab. Of course, in realistic situations material parameters are frequency dependent, but since the frequency variations in the response functions are determined by the permeability only and do not depend on the frequency explicitly, the resonant phenomena in subwavelength resonators suggested by Engheta are determined only by the resonant properties of the permeability function of the material layers. In this sense, phenomena in thin resonant layers resemble resonance in reflection from a thin ferrite layer on a metal plane or from a ferrite sphere near a microstrip line or in a closed waveguide.

If $\epsilon_2 = -\epsilon_1$ and $\mu_2 = -\mu_1$, the eigenvalue equations for the eigenwaves in a two-layer waveguide (6.155) and (6.156) or, as a special case, for surface waves along an interface between two media, are identically satisfied for arbitrary values of the propagation factor $\mathbf{k}_t$. This is a very special waveguide that supports waves with arbitrary wavenumbers at a fixed frequency. The most interesting case is when the waves between the planes are evanescent, meaning that $k_t^2 > |k_{1,2}|$. In this case the waves actually decay in the direction orthogonal to the interface, because the corresponding propagation factors $\beta_{1,2} = \sqrt{k_{1,2}^2 - k_t^2}$ are imaginary. Suppose that the volume between the two metal screens in Figure 6.30 is excited by a source with a certain fast variation of the current or field in space. Under the above assumptions regarding the media properties, all evanescent modes of this source satisfy the boundary conditions. Conceptually, this is an ideal memory device, because after the source has been removed, the field distribution near the interface of the two slabs will be preserved (until losses will consume the field energy), and the field distribution will correspond to the field distribution of the source.

This unique property wholly depends on the assumption that in one of the slabs the material parameters are negative. The slab thickness is irrelevant, and the same effect exists also on a single interface between two half-spaces filled by the same materials as the two slabs in the "resonator." Indeed, the memory effect for evanescent modes with arbitrary propagation constants is due to the fact that the interface supports surface modes with arbitrary propagation constants, at the frequency where $\mu_2 = -\mu_1$ and $\epsilon_2 = -\epsilon_1$. Actually, the same phenomenon is the core effect that makes the planar slab of a Veselago medium act as a perfect lens (Section 6.5.3). In that device, there are two such interfaces supporting surface waves with arbitrary wavenumbers. Resonant excitation of these modes leads to amplification of evanescent waves crossing the slab.

The main assumption has been that the Veselago material is an effective magnetodielectric *medium*, and the main challenge in realizing any device using the principle explained here is to design such a material with as small spatial period as possible. Higher-order evanescent modes vary extremely fast in space, and as soon as the spatial period of the exciting field becomes comparable with the spatial period of the artificial material with negative parameters, spatial dispersion effects degrade the properties of the device. Clearly, there are other limitations related to absorption present in any realistic medium and to the final size of the slabs in the transverse direction.

6.5.3   Perfect Lens

J.B. Pendry noticed [50] that a slab of an isotropic Veselago medium amplifies evanescent modes incident from one side of the slab, thus in principle making an image whose resolution is not restricted by the wavelength of the propagating waves. Indeed, let us consider a slab of thickness d (Figure 6.31) illuminated by an evanescent plane wave

$$E_y = e^{-jk_x x - \alpha_0 z}, \qquad H_x = -\frac{\alpha_0}{j\omega\mu_0} E_y \qquad (6.159)$$

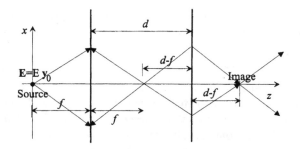

Figure 6.31 Planar slab of a Veselago medium with $\epsilon = -\epsilon_0$ and $\mu = -\mu_0$ illuminated by a TE-polarized wave (e.g., created by a current line along $\mathbf{y_0}$). Arrows show the direction of the phase vector. The same arrows show the distribution of the quasistatic electric field of a charged line near the slab, if we assume that the relative parameters equal to minus one at a very low frequency.

where $\alpha_0 = \sqrt{k_x^2 - k_0^2} > 0$, and $k_0 = \omega\sqrt{\epsilon_0\mu_0}$ (this is one Fourier component of the plane-wave spectrum created by the source). Looking for the field solution inside the slab as a sum of two evanescent plane waves

$$E_y = \left(Ae^{-\alpha z} + Be^{\alpha z}\right), \qquad H_x = -\frac{\alpha}{j\omega\mu}\left(Ae^{-\alpha z} - Be^{\alpha z}\right) \qquad (6.160)$$

where $\alpha = \sqrt{k_x^2 - k^2} > 0$, $k^2 = \omega^2\epsilon\mu$, and applying the boundary conditions on the two interfaces, the reflection and transmission coefficients can be found in the usual way. The result is

$$R = \frac{\frac{1}{2}\left(\frac{\alpha_0\mu}{\alpha\mu_0} - \frac{\alpha\mu_0}{\alpha_0\mu}\right)\sinh\alpha d}{\cosh\alpha d + \frac{1}{2}\left(\frac{\alpha_0\mu}{\alpha\mu_0} + \frac{\alpha\mu_0}{\alpha_0\mu}\right)\sinh\alpha d} \qquad (6.161)$$

$$T = \frac{1}{\cosh\alpha d + \frac{1}{2}\left(\frac{\alpha_0\mu}{\alpha\mu_0} + \frac{\alpha\mu_0}{\alpha_0\mu}\right)\sinh\alpha d} \qquad (6.162)$$

In the ideal case when inside the slab $\epsilon = -\epsilon_0$ and $\mu = -\mu_0$, the reflection coefficient vanishes and the transmission coefficient T becomes

$$T = e^{\alpha d} \qquad (6.163)$$

This tells us that every spatial harmonic of the field created by the source will be attenuated in air (the total attenuation factor is $e^{-\alpha[f+(d-f)]} = e^{-\alpha d}$), but *amplified* in the lens with the same rate, thus theoretically realizing the perfect lens.

For better understanding of this interesting and strange phenomenon of "field amplification," let us turn to a simpler system of only one interface between free space and an isotropic Veselago medium. Let this interface be excited by a plane propagating or evanescent wave. Assuming the incident field in form (6.159), and the transmitted field as

$$E_y^{\text{trans}} = T \cdot E_y^{\text{inc}} e^{-jk_x x - \alpha z}, \qquad H_x^{\text{trans}} = -\frac{\alpha}{j\omega\mu} E_y^{\text{trans}} \qquad (6.164)$$

we can readily find the reflection and transmission coefficients as

$$R = \frac{1 - \frac{\alpha\mu_0}{\alpha_0\mu}}{1 + \frac{\alpha\mu_0}{\alpha_0\mu}}, \qquad T = \frac{2}{1 + \frac{\alpha\mu_0}{\alpha_0\mu}} \qquad (6.165)$$

If the second medium is characterized by usual positive material parameters, then for $\mu \to \mu_0$ and $\epsilon \to \epsilon_0$ (meaning $\alpha \to \alpha_0$) the transmission coefficient naturally becomes equal to 1, and the reflection coefficient vanishes for all plane waves, propagating or evanescent. However, if the material parameters of the second medium are both negative, the result is dramatically different. If the wave is a propagating wave (that is, α is imaginary), then with the change of the sign of μ we must also change the sign of the phase constant α: This is a *backward-wave* medium, and the Poynting vector must be directed from the source. But if the incident wave is an evanescent mode (real α), the sign of the *decay* factor is the same for positive or negative μ, because the field must in any case decay from the source. Thus, the interface in the case $\mu \to -\mu_0$ and $\epsilon \to -\epsilon_0$ is transparent for propagating modes, but for evanescent modes both R and T become infinite.

The reason for this behavior is the fact that for a fixed frequency, the interface supports surface waves with arbitrary propagation constants along the interface. This is clear from (6.156): We have already seen that in the limit $d_{1,2} \to \infty$ with $\mu \to -\mu_0$ and $\epsilon \to -\epsilon_0$, the eigenvalue equation is satisfied identically for all k_t. What actually happens is that any evanescent incident plane wave is exactly in phase with one of the eigenmodes of the surface wave spectrum. As the interface is infinite and we have neglected losses, the amplitude of the excited surface wave becomes infinite. The physics of the evanescent field amplification in Veselago slabs is very similar. The incident field excites an eigenmode of the slab that is formed by two exponentially decaying field components inside the slab. The spectrum of eigenmodes can be found by equating the denominator of (6.161) or (6.162) to zero. For material parameters satisfying $\mu = -\mu_0$ and $\epsilon = -\epsilon_0$ the denominator equals simply $\exp(-\alpha d)$. The larger the value of α, the closer the excitation is to the resonance with the waveguide eigenmode, and the larger the field amplitude excited in the slab waveguide. As in excitations of any high-quality resonators, the time needed to develop oscillations increases with increasing the quality factor.

As an example, let the lens be excited by a current line (as in Figure 6.31). The incident electric field is proportional to $H_0^{(2)}(k\sqrt{x^2 + z^2})$, whose Fourier transform is (4.12)

$$\int_{-\infty}^{\infty} H_0^{(2)}(k\sqrt{x^2 + z^2})e^{-jk_x x}\,dx = \frac{2}{\sqrt{k^2 - k_x^2}}e^{-j\sqrt{k^2 - k_x^2}|z|} \qquad (6.166)$$

For simplicity, if the source location is just at the lens surface ($z = 0$), the amplitudes of the spatial harmonics are $2/\sqrt{k^2 - k_x^2}$. On the other side of the lens the propagating waves become

$$\frac{2}{\sqrt{k^2 - k_x^2}}e^{+j\sqrt{k^2 - k_x^2}d}, \qquad k_x < k \qquad (6.167)$$

but the evanescent part of the spectrum transforms like

$$\frac{2}{\sqrt{k^2 - k_x^2}}e^{\sqrt{k_x^2 - k^2}d}, \qquad k_x > k \qquad (6.168)$$

Next, propagating to the focal point, the components transform as

$$\frac{2}{\sqrt{k^2 - k_x^2}}e^{+j\sqrt{k^2 - k_x^2}d}e^{-j\sqrt{k^2 - k_x^2}d} = \frac{2}{\sqrt{k^2 - k_x^2}}, \qquad k_x < k \qquad (6.169)$$

$$\frac{2}{\sqrt{k^2 - k_x^2}}e^{\sqrt{k_x^2 - k^2}d}e^{-\sqrt{k_x^2 - k^2}d} = \frac{2}{\sqrt{k^2 - k_x^2}}, \qquad k_x > k \qquad (6.170)$$

and we see that indeed this is the spectrum of the original source. By the way, note that the image is transformed without any phase advance. Now let us look at the field spectrum just at the second surface of the lens ($z = d$) given by (6.167) and (6.168). If we want to know how the field is distributed over this surface, we should calculate the inverse Fourier transform of this spectrum. Here we arrive at a serious difficulty [51], since the corresponding integral has no absolute convergence [the spectrum given by (6.168) exponentially grows with increasing $|k_x|$]. In reality, the lens material is discrete (more on Veselago media realizations in Section 6.5.5), so the lens cannot amplify too quickly varying evanescent components. This sets a limit for the meaningful values of $|k_x|$ as well as for the lens performance.

Several other factors will severely restrict any attempt to realize a perfect lens with the use of materials with negative parameters. First, the field amplification is limited by inevitable losses in the material [lossy lens can be easily analyzed using (6.161) and (6.162) with complex material parameters]. The second factor originates from the fact that the lens operation is based on excitation of surface waves traveling along the lens surface. Clearly, the finite size of the slab in the transverse plane will cause reflections from the lens border that will strongly affect the lens effect.

It is interesting that for the amplification of evanescent modes it is not critical that the medium inside the slab be a backward-wave medium[14] in the sense that the evanescent modes decay in this medium in exactly the same way as in a usual material. What is critical is that the two interfaces are *resonant* and support surface waves with large wavenumbers. If it would be possible to realize such surfaces by other means, at least for one or several fixed wavenumbers, the system would amplify evanescent waves with these propagation constants even if there would be free space between the two surfaces. Actually, the mechanism of this amplification is the same as the excitation of high-amplitude oscillations in a high-quality resonator by a weak but resonant source.

6.5.4   Electromagnetic Crystals with Veselago Layers

Let us consider an infinite periodic structure composed of alternating layers of two materials with different relative permittivities ϵ_1, ϵ_2 and permeabilities μ_1, μ_2 [52]. Focusing on the spatial resonances in the structure, we assume here that the material parameters do not depend on the frequency. As shown in Figure 6.32, one of the layers is a usual isotropic and lossless material ($\epsilon_1 > 0$, $\mu_1 > 0$), but the other layer in every period is made of a Veselago medium ($\epsilon_2 < 0$, $\mu_2 < 0$). The thicknesses of the two layers are $d_{1,2}$, respectively, and the period is denoted by $L = d_1 + d_2$. The propagation constant β of eigenwaves in this periodical structure can be found from the well-known eigenvalue equation, whose form is the same as for the corresponding structure made of usual materials (5.134):

$$\cos \beta L = \cos(k_1 d_1) \cos(k_2 d_2) - \frac{\eta_1^2 + \eta_2^2}{2\eta_1 \eta_2} \sin(k_1 d_1) \sin(\pm k_2 d_2) \qquad (6.171)$$

Here $\eta_i = \sqrt{\mu_i/\epsilon_i}$ ($i = 1, 2$), $k_i = k n_i$, $n_i = \sqrt{\epsilon_i \mu_i}$, are the absolute values of the refractive indices of the two layers forming every period, and k is the wavenumber in vacuum. The signs of the material parameters and of the refractive indices are explicitly included in this formula. The only difference in the derivation of (6.171) for the new structure is that in the slabs of negative materials the sign of the phase constant k_2 must be reversed. This corresponds to the lower sign in the last term of (6.171). The upper sign gives the equation for the usual case when $\epsilon_2 > 0$, $\mu_2 > 0$.

If the thicknesses of the two layers forming each period are the same, the finite-period structure has well-pronounced bandgaps with the centers corresponding to the wavelengths, satisfying the quarter-wave condition for the first bandgap (and 3/4, 5/4, and so on for the other stopbands), as also takes place in bandgap structures composed of usual materials. Figure 6.33

---

[14]Although negative refraction is necessary to focus the *propagating* part of the Fourier spectrum.

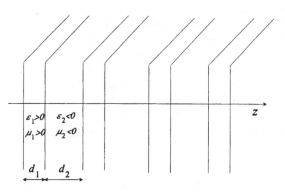

Figure 6.32 One-dimensional electromagnetic crystal with intermitting layers of a usual material and a Veselago material.

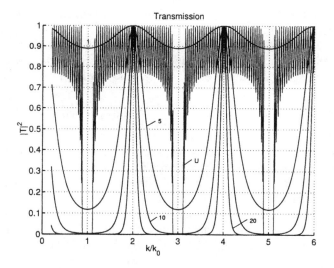

Figure 6.33 Transmission coefficient through a finite-length one-dimensional electromagnetic crystal (normal incidence). Crystal formed by two usual materials (curve marked by "U") and a crystal with Veselago slabs are compared. Note how the extremely wide bandgaps separated by narrow passbands are formed when the number of layers (shown by numbers) increases.

demonstrates the spectral transmission of a structure with Veselago slabs for different numbers of periods. The transmittance of a usual 20-period structure is presented for comparison (marked by "U"). When the number of periods increases, very wide stopbands are formed, separated by very narrow transmission windows at wavelengths $2d_1n_1$, d_1n_1, and so forth ($k/k_0 = 2, 4, \ldots$). At other frequencies the structure acts as a reflector.

6.5.5 Possible Realizations of Media with Negative Parameters

A two-phase composite was used in the first reported realization of Veselago media [42], so that the electric and magnetic responses were provided by different subsystems. Negative permittivity was due to a periodical array of thin parallel conducting wires; see the model in Chapter 5, Section 5.5. Magnetic response was generated by an array of double split-ring resonators. We can understand their properties using the theory of loaded loop scatterers (Section 5.1.2) and the theory of double spiral scatterers [53]. Double split rings, like single loaded loops, are bianisotropic particles [6, 54–56]. In other words, an external electric field produces a magnetic dipole in the particle and, at the same time, an external magnetic field induces an electric dipole. For double split-ring resonators, this magnetoelectric effect is reduced by choosing the opposite positioning of the slits of both loops. However, it cannot cancel out exactly if the loops are of different sizes.

Realizations as Composites with Resonant Magnetoelectric Inclusions

If the electric and magnetic responses are provided by different sets of inclusions, like long wires and split rings, the frequency dispersion of the effective permittivity and permeability is very different. However, it appears to be possible to generate both polarizations using inclusions of only one shape. Clearly, they must be polarizable by both electric and magnetic fields. Such particles are well known in the design of reciprocal chiral and other bianisotropic composite materials [4]. Basically, the inclusion should contain a loop portion, to react on magnetic fields, and straight-wire portion, to react on electric field. One possible realization can be a racemic[15] mixture or an array of canonical helices (see Figure 6.34).

In Figure 6.35, an example for the material parameters of a racemic chiral composite medium is shown (calculated with the use of a simplified analytical model).[16] The left-hand side is for a wide frequency range, and

---

[15]A mixture of an equal number of right- and left-handed chiral particles, so that the net effective chirality of the material is zero.

[16]These calculations have been made by S.I. Maslovski.

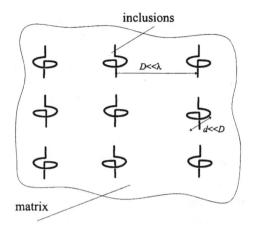

Figure 6.34 The geometry of a racemic composite. The number of right- and left-handed particles is the same in every unit volume.

the right one is a blowup in a narrower region where the real parts of the parameters are both negative. The inclusion dimensions and the concentration have been chosen so that the real parts of the permittivity and permeability are approximately equal. The medium inclusions are the canonical chiral particles [5, 57] resonating near 10 GHz. The metal inclusions are made of silver. Absorption in the particles leads to nonzero imaginary parts of the permittivity and permeability (see Figure 6.35). The arrangement of inclusions is supposed to be regular and dense, so we can neglect the scattering losses in the composite.

Let us now consider the reflection and transmission for a slab of such material. These are presented in Figure 6.36, for a 3-cm-thick slab. We can

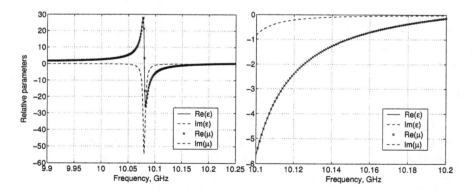

Figure 6.35 The effective material parameters of a racemic composite.

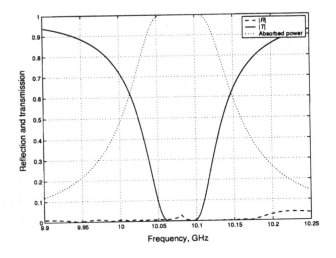

Figure 6.36 Reflection and transmission through a racemic slab with negative parameters near 10 GHz.

see that in spite of the absence of scattering losses, the resistive losses are high at the particle resonance. The wave impedance of the effective medium is nearly the same as that of free space and reflection is almost zero everywhere. The transmission coefficient turns out to be very small within the resonance band. However, the transmitted wave appears to be measurable in the region of negative parameters. At 10.15 GHz the real parts of both the permittivity and permeability are close to -1, and the transmittance is about -3 dB.

Wire Media and Resonant Magnetic Particles Combined[17]

The effective permittivity of wire media formed by wires in free space is given by (5.210). If between cylinders there is a certain isotropic medium characterized by relative permittivity ϵ and relative permeability μ, replace ϵ_0 by $\epsilon\epsilon_0$ and μ_0 by $\mu\mu_0$:

$$\overline{\overline{\epsilon}}_{\text{eff}} = \epsilon_0 \left[\epsilon\overline{\overline{I}} - \frac{2\pi \mathbf{z}_0 \mathbf{z}_0}{\mu(k_0 a)^2 \log \frac{a^2}{4r_0(a - r_0)}} \right] \tag{6.172}$$

k_0 is still the free-space wavenumber here. We can see from (6.172) that if the filling medium has usual properties ($\mu > 0$, $\epsilon > 0$), the negative values of ϵ_{eff} are possible.

---

[17]This subject has been studied by S.I. Maslovski, whose results [58] are used in the section.

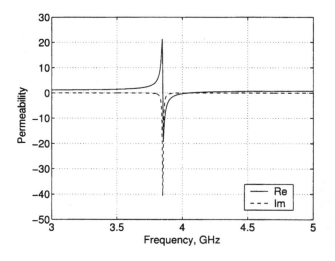

Figure 6.37 Typical relative effective permeability of a mixture of split-ring resonators as a function of the frequency.

Let us suppose that the wires are immersed into a negative permeability medium, that is, $\epsilon > 0$ but $\mu < 0$. We see that it is impossible to realize negative values of the mixture permittivity in this medium. This conclusion is quite general, assuming that there is indeed a *continuous medium* between the cylinders. As an example, let us consider a system composed of split-ring resonators resonating at approximately 3.85 GHz. The resonators are placed inside a wire medium whose plasma frequency (at which the effective permittivity changes sign) is about 10 GHz. A simple *LC*-circuit model and the Maxwell Garnett mixing rule have been used to model the split-ring subsystem. Scattering losses are assumed to be compensated by the particle interactions, as in regular crystals. Equation (6.172) has been used to model the cylinder array forming the wire-medium subsystem. Losses in the materials (copper wires) are taken into account. Figure 6.37 shows the permeability of the split-ring subsystem. The permeability of the entire system is the same.

Figure 6.38, left, shows the permittivity of the wire medium without magnetic inclusions. This result is well known (see Section 5.5.4). The whole system (split-ring array plus wire medium) has such permittivity when the split rings are placed at the symmetry planes where the magnetic field of wire currents is zero and the split rings do not feel the presence of long wires.

For the situation when the quasistatic interaction between split rings and wires does not vanish (e.g., a wire array immersed in a random mixture of small split-ring particles), the effective permittivity is presented on the right-hand side of Figure 6.38. It is seen that the real part of the medium

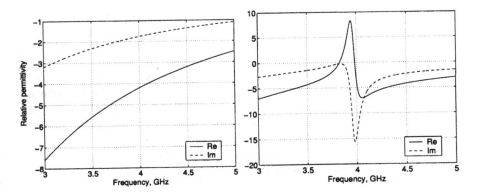

Figure 6.38 Effective permittivity of a system containing an array of long metal cylinders and a mixture of split-ring resonators. Left: no magnetic inclusions.

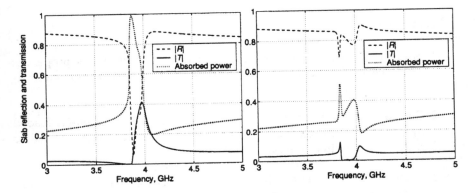

Figure 6.39 Slab reflection and transmission. Left: no magnetic field interaction between the two subsystems. Right: random space distribution of split-ring resonators.

permittivity becomes positive in the region of strong resonant behavior of the magnetic subsystem. Two plots on Figure 6.39 show simulated transmission and reflection coefficients for a 2-cm slab of this material: The left plot is for the arrangement as in [41, 42], and the right one is for the case of strong interaction between the two subsystems. Also, the absorption power density in the slabs is depicted on the plots by dotted lines. We can see that the transparency peak nearly disappears on the second plot.

Realizations as Active Composites: Broadband Response

The general concept of metamaterials based on composites with small loaded dipole antennas (Section 6.1.1) can be in principle applied in the design of materials with negative parameters [10]. Let us demand that $\epsilon_{\text{eff}} = -|\epsilon_r|\epsilon_0$ and $\mu_{\text{eff}} = -|\mu_r|\mu_0$, where $\epsilon_r < 0$ and $\mu_r < 0$. If we require the effective parameters be independent from the frequency,[18] the loads for the wire dipole inclusions must be capacitances, and (6.7) gives for the load capacitance

$$C_{\text{load}} = -\frac{\frac{Nl^2 C_{\text{wire}}}{3} + (|\epsilon_r| + 1)\frac{3\epsilon_0}{2 - |\epsilon_r|}}{\frac{4Nl^2 C_{\text{wire}}}{3} + (|\epsilon_r| + 1)\frac{3\epsilon_0}{2 - |\epsilon_r|}} C_{\text{wire}} \qquad (6.173)$$

that is in general negative. Similarly, (6.11) gives for the inductance of the loads connected to the loop inclusions

$$L_{\text{load}} = -L_{\text{loop}} + NS^2 \mu_0 \frac{2 - |\mu_r|}{3(|\mu_r| + 1)} \qquad (6.174)$$

PROBLEMS

6.1 Study the possibility of designing a metamaterial with the properties of the perfectly matched layer (PML) so that at least one of the loads of the artificial molecules is a short circuit.

6.2 Consider a planar regular array of small wire dipole antennas loaded by arbitrary bulk loads and determine the load impedance needed to realize the required reflection coefficient.

6.3 Prove that the uniaxial PML material absorbs power of an arbitrary plane wave traveling in the medium.

6.4 Develop a model for an interface between wire medium and free space such that the wires are orthogonal to the interface. Hint: Charge accumulation at the wire ends will lead to an additional capacitance that can be described in terms of an equivalent thin dielectric layer between the wire medium and free space.

6.5 Consider possible designs of metamaterials with negative parameters using loaded dipoles (Section 6.5.5). Study what parameter ranges require negative or positive capacitive and inductive loads and discuss what difficulties to expect in a practical realization of the composite.

---

[18] Frequency-independent negative material parameters mean that the energy density is negative, meaning that a sample of this material is a source of energy.

References

[1] Sihvola, A., "Electromagnetic Emergence in Metamaterials. Deconstraction of Terminology of Complex Media," Zouhdi, S., A. Sihvola, and M. Arsalane, (eds.), *Advances in Electromagnetics of Complex Media and Metamaterials*, Dordrecht, the Netherlands: Kluwer Academic Publishers, 2003, pp. 1-18.

[2] Pearsall, J., (ed.), *The New Oxford Dictionary of English*, Oxford, England: Oxford University Press, 1998.

[3] Mariotte, F., B. Sauviac, and S.A. Tretyakov, "Artificial Bi-Anisotropic Composites," Werner, D.H., and R. Mittra, (eds.), *Frontiers in Electromagnetics*, New York: IEEE Press, 2000, pp. 732-770.

[4] Serdyukov, A.N., et al., *Electromagnetics of Bi-Anisotropic Materials: Theory and Applications*, Amsterdam: Gordon and Breach Science Publishers, 2001.

[5] Tretyakov, S.A., et al., "Analytical Antenna Model for Chiral Scatterers: Comparison with Numerical and Experimental Data," *IEEE Trans. Antennas and Propagation*, Vol. 44, No. 7, 1996, pp. 1006-1014.

[6] Simovski, C.R., et al., "Antenna Model for Conductive Omega Particles," *J. Electromagnetic Waves and Applications*, Vol. 11, No. 11, 1997, pp. 1509-1530.

[7] Auzanneau, F., and R.W. Ziolkowski, "Theoretical Study of Synthetic Bianisotropic Materials," *J. Electromagnetic Waves and Applications*, Vol. 12, No. 3, 1998, pp. 353-370.

[8] Auzanneau, F., and R.W. Ziolkowski, "Microwave Signal Rectification Using Artificial Composite Materials Composed of Diode-Loaded Electrically Small Dipole Antennas," *IEEE Trans. Microwave Theory and Techniques*, Vol. 46, No. 11, 1998, pp. 1628-1637.

[9] Tretyakov, S.A., and T.G. Kharina, "The Perfectly Matched Layer as a Synthetic Material with Active Inclusions," *Electromagnetics*, Vol. 20, No. 2, 2000, pp. 155-166.

[10] Tretyakov, S.A., "Meta-Materials with Wideband Negative Permittivity and Permeability," *Microwave and Optical Technology Lett.*, Vol. 31, No. 3, 2001, pp. 163-165.

[11] El Khoury, S., "The Design of Active Floating Positive and Negative Inductors in MMIC Technology," *IEEE Microwave and Guided Wave Lett.*, Vol. 5, No. 10, 1995, pp. 321-323.

[12] Taflove, A., and S.C. Hagness, *Computational Electrodynamics: The Finite-Difference Time-Domain Method,* 2nd ed., Norwood, MA: Artech House, 2000.

[13] Berenger, J.-P., "A Perfectly Matched Layer for the Absorption of Electromagnetic Waves," *J. Computational Physics,* Vol. 114, 1994, pp. 185-200.

[14] Gedney, S.D., "An Anisotropic Perfectly Matched Layer – Absorbing Medium for the Truncation of FDTD Lattices," *IEEE Trans. Antennas and Propagation,* Vol. 44, No. 12, 1996, pp. 1630-1639.

[15] Lindell, I.V., *Methods for Electromagnetic Field Analysis,* Oxford, England: Clarendon Press, 1992.

[16] Ziolkowski, R.W., "The Design of Maxwellian Absorbers for Numerical Boundary Conditions and for Practical Applications Using Engineered Artificial Materials," *IEEE Trans. Antennas and Propagation,* Vol. 45, No. 4, 1997, pp. 656-671.

[17] Tretyakov, S.A., "Uniaxial Omega Medium as a Physically Realizable Alternative for the Perfectly Matched Layer (PML)," *J. Electromagnetic Waves and Applications,* Vol. 12, 1998, pp. 821-837.

[18] Rozanov, K.N., "Ultimate Thickness to Bandwidth Ratio of Radar Absorbers," *IEEE Trans. Antennas and Propagation,* Vol. 48, No. 8, 2000, pp. 1230-1234.

[19] Belov, P.A., and S.A. Tretyakov, "Resonant Reflection from Dipole Arrays Located Very Near to Conducting Planes," *J. Electromagnetic Waves and Applications,* Vol. 16, No. 1, 2002, pp. 129-143.

[20] Lourtioz, J.-M., et al., "Toward Controllable Photonic Crystals for Centimeter- and Millimeter-Wave Devices," *J. Lightwave Technology,* Vol. 18, 1999, No. 11, pp. 2025-2031.

[21] Simovski, C.R., and S. He, "Antennas Based on Modified Metallic Photonic Bandgap Structures Consisting of Capacitively Loaded Wires," *Microwave and Optical Technology Lett.,* Vol. 31, No. 3, 2001, pp. 214-221.

[22] Smith, D.R., et al., "Loop-Wire Medium for Investigating Plasmons at Microwave Frequencies," *Applied Physics Lett.,* Vol. 75, No. 10, 1999, pp. 1425-1427.

[23] Belov, P.A., C.R. Simovski, and S.A. Tretyakov, "Two-Dimensional Electromagnetic Crystals Formed by Reactively Loaded Wires," *Physical Review E,* Vol. 66, 2002, p. 036610.

[24] Belov, P.A., S.A. Tretyakov, and A.J. Viitanen, "Nonreciprocal Microwave Bandgap Structures," *Physical Review E*, Vol. 66, 2002, p. 016608.

[25] Maslovski, S.I, S.A. Tretyakov, and P.A. Belov, "Wire Media with Negative Effective Permittivity: A Quasi-Static Model," *Microwave and Optical Technology Lett.*, Vol. 35, No. 1, 2002, pp. 47-51.

[26] Sievenpiper, D., et al., "High-Impedance Electromagnetic Surfaces with a Forbidden Frequency Band," *IEEE Trans. Microwave Theory Techniques*, Vol. 47, No. 11, 1999, pp. 2059-2074.

[27] Sievenpiper, D.F., "High-Impedance Electromagnetic Surfaces," Ph.D. thesis, University of California at Los Angeles, 1999, available at http://www.ee.ucla.edu/labs/photon/thesis/ThesisDan.pdf

[28] Yang, F.-R., et al., "A Novel TEM Waveguide Using Uniplanar Compact Photonic-Bandgap (UC-PBG) Structure," *IEEE Trans. Microwave Theory Techniques*, Vol. 47, No. 11, 1999, pp. 2092-2098.

[29] Tretyakov, S.A., and C.R. Simovski, "Wire Antenna Near Artificial Impedance Surface," *Microwave and Optical Technology Lett.*, Vol. 27, No. 1, 2000, pp. 46-50.

[30] Diaz, R.E., J.T. Aberle, and W.E. McKinzie, "TM Mode Analysis of a Sievenpiper High-Impedance Reactive Surface," *Antennas and Propagation Society International Symposium*, Vol. 1, 2000, pp. 327-330.

[31] Tretyakov, S.A., and C.R. Simovski, "Dynamic Model of Artificial Reactive Impedance Surfaces," *J. Electromagnetic Waves and Applications*, Vol. 17, No. 1, 2003, pp. 131-145.

[32] Kong, J.A., *Electromagnetic Wave Theory*, New York: John Wiley & Sons, 1986.

[33] Compton, R.C., L.B. Whitbourn, and R.C. McPhedran, "Strip Gratings at a Dielectric Interface and Application of Babinet's Principle," *Applied Optics*, Vol. 23, No. 18, 1984, pp. 3236-3242.

[34] Whitbourn, L.B., and R.C. Compton, "Equivalent-Circuit Formulas for Metal Grid Reflectors at a Dielectric Boundary," *Applied Optics*, Vol. 24, No. 2, 1985, pp. 217-220.

[35] Yatsenko, V.V., et al., "Higher Order Impedance Boundary Conditions for Sparse Wire Grids," *IEEE Trans. Antennas and Propagation*, Vol. 48, No. 5, 2000, pp. 720-727.

[36] Sochava, A.A., "Diffraction of Electromagnetic Waves by a Semi-Infinite Grid Positioned at the Earth Surface," M.Sc. thesis, Leningrad Polytechnic Institute, 1987 (in Russian).

[37] Wait, J.R., "Reflection from a Wire Grid Parallel to a Conducting Plane," *Can. J. Physics,* Vol. 32, 1954, pp. 571-579.

[38] Saville, M.A., "Investigation of Conformal High-Impedance Ground Planes," Thesis AFIT/GE/ENG/00M-17, Air Force Institute of Technology, Wright-Patterson Air Force Base, Ohio, available at http://www.au.af.mil/au/database/research/ay2000/afit/afit-ge-eng-00m-17.htm

[39] Felsen, L.B., and N. Marcuvitz, *Radiation and Scattering of Waves,* Piscataway, NJ: IEEE Press, 1994.

[40] Veselago, V.G., "The Electrodynamics of Substances with Simultaneously Negative Values of ϵ and μ," *Soviet Physics Uspekhi,* Vol. 10, 1968, pp. 509-514 (originally pubished in Russian in *Uspekhi Fizicheskikh Nauk,* Vol. 92, 1967, pp. 517-526).

[41] Smith, D.R., et al., "Composite Media with Simultaneously Negative Permeability and Permittivity," *Physical Review Lett.,* Vol. 84, 2000, pp. 4184-4187.

[42] Shelby, R.A., D.R. Smith, and S. Schultz, "Experimental Verification of a Negative Index of Refraction," *Science,* Vol. 292, 2001, pp. 77-79.

[43] Pendry, J.B., et al., "Magnetism from Conductors and Enhanced Nonlinear Phenomena," *IEEE Trans. Microwave Theory and Techniques,* Vol. 47, 1999, pp. 2075-2084.

[44] Valanju, P.M., R.M. Walser, and A.P. Valanju, "Wave Refraction in Negative-Index Media: Always Positive and Very Inhomogeneous," *Physical Review Lett.,* Vol. 88, 2002, p. 187401.

[45] Smith, D.R., D. Schurig, and J.B. Pendry, "Negative Refraction of Modulated Electromagnetic Waves," *Applied Phys. Lett.,* Vol. 81, No. 5, 2002, pp. 2713-2715.

[46] Pacheco, J., Jr., T.M. Grzegorczyk, B.-I. Wu, Y. Zhang, and J.A. Kong, "Power Propagation in Homogeneous Isotropic Frequency-Dispersive Left-Handed Media," *Physical Review Lett.,* Vol. 89, No. 25, 2002, p. 257401.

[47] Pendry, J.B., and D.R. Smith, "Comment on 'Wave Refraction in Negative-Index Media: Always Positive and Very Inhomogeneous,'" *Physical Review Lett.,* Vol. 90, 2003, p. 029703.

[48] Landau, L.D., E.M. Lifshitz, and L.P. Pitaevskii, *Electrodynamics of Continuous Media*, 2nd ed., Oxford, England: Butterworth-Heinemann, 1984.

[49] Engheta, N., "An Idea for Thin Subwavelength Cavity Resonators Using Metamaterials with Negative Permittivity and Permeability," *IEEE Antennas and Wireless Propagation Lett.*, Vol. 1, No. 1, 2002, pp. 10-13.

[50] Pendry, J.B., "Negative Refraction Makes a Perfect Lens," *Physical Review Lett.*, Vol. 85, 2000, pp. 3966-3969.

[51] Garcia, N., and M. Nieto-Vesperinas, "Left-Handed Materials Do Not Make a Perfect Lens," *Physical Review Lett.*, Vol. 88, 2002, p. 207403.

[52] Nefedov, I.S., and S.A. Tretyakov, "Photonic Band Gap Structure Containing Metamaterial with Negative Permittivity and Permeability," *Physical Review E*, Vol. 66, 2002, p. 036611.

[53] Lagarkov, A.N., et al., "Resonance Properties of Bi-Helix Media at Microwaves," *Electromagnetics*, Vol. 17, No. 3, 1997, pp. 213-237.

[54] Marques, R., F. Medina, and R.R. El-Idrissi, "Role of Bianisotropy in Negative Permeability and Left-Handed Metamaterials," *Physical Review B*, Vol. 65, 2002, p. 14440.

[55] Saadoum, M.M.I., and N. Engheta, "A Reciprocal Phase Shifter Using Novel Pseudochiral or Omega Medium," *Microwave and Optical Technology Lett.*, Vol. 5, 1992, pp. 184-188.

[56] Kharina, T.G., et al., "Experimental Study of Artificial Omega Media," *Electromagnetics*, Vol. 18, 1998, pp. 437-457.

[57] Jaggard, D.L., A.R. Mickelson, and C.T. Papas, "On Electromagnetic Waves in Chiral Media," *Applied Physics*, Vol. 18, 1979, pp. 211-216.

[58] Maslovski, S.I., "On the Possibility of Creating Artificial Media Simultaneously Possessing Negative Permittivity and Permeability," *Technical Physics Lett.*, Vol. 29, No. 1, 2003, pp. 32-34.

List of Symbols

Vectors are denoted by bold letters (**a**); dyadics are written as $\overline{\overline{A}} = \mathbf{ab} + \mathbf{cd} + \ldots$. The scalar product of two vectors is denoted by dot ($\mathbf{a} \cdot \mathbf{b}$), and the vector product is denoted by cross ($\mathbf{a} \times \mathbf{b}$). Time-harmonic dependence is in the form $\exp(j\omega t)$.

a grid or lattice period length; loop radius

B   magnetic induction

$c = 1/\sqrt{\epsilon_0 \mu_0}$ the speed of light in vacuum

C capacitance; interaction constant in three-dimensional arrays or normalized interaction constant

d layer thickness

D   displacement vector

E   electric field

E_n normal component of the electric field

$\mathbf{E}_t$ tangential electric field component

$\mathbf{E}_{t\pm}$ tangential electric field component at the two opposite sides of a slab

$\mathbf{E}^{\text{ext}} = \mathbf{E}^{\text{inc}}$ external or incident electric field

$\mathbf{E}^{\text{loc}}$ local electric field

$\mathbf{E}^{\text{tot}}$ total electric field

H   magnetic field

H_n normal component of the magnetic field

$\mathbf{H}_t$ tangential magnetic field component

$\mathbf{H}_{t\pm}$ tangential magnetic field component at the two opposite sides of a slab

I electric current

$\overline{\overline{I}} = \mathbf{x}_0\mathbf{x}_0 + \mathbf{y}_0\mathbf{y}_0 + \mathbf{z}_0\mathbf{z}_0$ the unit dyadic

$\overline{\overline{I}}_t = \overline{\overline{I}} - \mathbf{z}_0\mathbf{z}_0$ transverse (with respect to axis z) unit dyadic

j imaginary unit

k   wave vector in medium

$\mathbf{k}_t$ tangential component of the wave vector

$k_0 = \omega\sqrt{\epsilon_0 \mu_0}$ free-space wavenumber

l arm length of a dipole antenna

L inductance

m   magnetic dipole moment

n   unit vector normal to a surface or an interface

N inclusion concentration (number of inclusions per unit volume)

p   electric dipole moment

q propagation constant

R reflection coefficient

r_0 wire radius

t time

T transmission coefficient

U electric voltage

x$_0$ unit vector along Cartesian axis x

$\overline{\overline{Y}} = \overline{\overline{Z}}^{-1}$, $Y = 1/Z$ dyadic or scalar admittance

y$_0$ unit vector along Cartesian axis y

z$_0$ unit vector along Cartesian axis z

$\overline{\overline{Z}} = \overline{\overline{Y}}^{-1}$, $Z = 1/Y$ dyadic or scalar impedance

Z_g grid impedance

Z_{inp} input impedance (of an antenna)

Z_s surface impedance

α grid parameter

$\overline{\overline{\alpha}}, \alpha$ dyadic or scalar polarizability

$\beta = \sqrt{k^2 - k_t^2}$ the axial or normal component of the propagation factor; interaction constant in two-dimensional arrays

ϵ permittivity

ϵ_0 permittivity of vacuum

ϵ_{eff} effective permittivity of a composite medium

$\epsilon' = \text{Re}\{\epsilon\}$ the real part of the permittivity

$\epsilon'' = \text{Im}\{\epsilon\}$ the imaginary part of the permittivity

$\eta = \sqrt{\mu/\epsilon}$ wave impedance of an isotropic medium

$\eta_0 = \sqrt{\mu_0/\epsilon_0}$ wave impedance of plane waves in vacuum

θ incidence angle

λ wavelength

μ permeability

μ_0 permeability of vacuum

ω angular frequency

$\nabla_t = \nabla - \frac{\partial}{\partial z}\mathbf{z}_0\mathbf{z}_0$ two-dimensional gradient operator

About the Author

Sergei Tretyakov received Dipl. Engineer-Physicist, Ph.D., and Doctor of Sciences degrees (all in radiophysics) from the St. Petersburg State Technical University in Russia, in 1980, 1987, and 1995, respectively. From 1980 to 2000 Professor Tretyakov worked for the Radiophysics Department of the St. Petersburg State Technical University. From 1993 to 1999, he was the director of the Complex Media Electromagnetics Laboratory of that university. He has also been a visiting scientist in the Electromagnetics Laboratory of the Helsinki University of Technology. In 1994, he was a visiting scientist with the French Atomic Energy Commission (CEA-CESTA), Le Barp, France, and also affiliated with the University of Bordeaux, Laboratory of Wave-Material Interactions (PIOM Laboratory).

Presently, Sergei Tretyakov is a professor of radio engineering in the Radio Laboratory at the Helsinki University of Technology. His main scientific interests are electromagnetic field theory, complex media electromagnetics, and microwave engineering. Professor Tretyakov has been an associate editor of *Electromagnetics, Journal of Electromagnetic Waves and Applications,* and *Radio Science.* He organized the St. Petersburg IEEE ED/MTT/AP Chapter and served as its chairman from 1995 to 1998.

Index

Recent Titles in the Artech House Electromagnetic Analysis Series

Tapan K. Sarkar, Series Editor

Understanding Electromagnetic Scattering Using the Moment Method: A Practical Approach, Randy Bancroft

Wavelet Applications in Engineering Electromagnetics, Tapan K. Sarkar, Magdalena Salazar-Palma, and Michael C. Wicks

For further information on these and other Artech House titles, including previously considered out-of-print books now available through our In-Print-Forever® (IPF®) program, contact:

Artech House
685 Canton Street
Norwood, MA 02062
Phone: 781-769-9750
Fax: 781-769-6334
e-mail: artech@artechhouse.com

Artech House
46 Gillingham Street
London SW1V 1AH UK
Phone: +44 (0)20 7596-8750
Fax: +44 (0)20 7630 0166
e-mail: artech-uk@artechhouse.com

Find us on the World Wide Web at:
www.artechhouse.com